10 Days To Multiplication Mastery

and more

(A Commutative Approach)

Written by
Marion W. Stuart

Illustrated by
R. Matthew Stuart

Sincere appreciation:

To Ron Stuart for his insistent persuasion to publish this workshop-and-classroom proven material into one easy-to-use, hands-on book.

To Lynnette W. Hancock for her encouragement and hours of editing.

To artists R. Matthew Stuart, Kelly Taylor (the cover), Val C. Bagley (the Counting Cards and art ideas), and Jim Aylesworth (Finger 9's).

The staff at Learning Wrap-ups, and my precious family who each helped in their own way.

To Susan Myers for her editing suggestions and help with the 1996 revision.

A SPECIAL THANKS TO THE MANY TEACHERS AND STUDENTS WHO CONTRIBUTED IDEAS AND USED ALL OR PART OF THESE MATERIALS AND PROVED THEM SUCCESSFUL.

© Learning Wrap-ups Inc.
Printings 1995, 1996, 1997
Ogden, Utah 84405

ISBN # 0-943343-69-0   10 Days to Multiplication Mastery and More (a Commutative Approach)

CONTENTS

SECTION ONE

GETTING STARTED
(Teacher information)

INTRODUCTION

Two great teaching concepts have come together to make the 10 DAYS program work.

1. LEARNING WRAP-UP DRILL, which enables students to learn math facts at amazing speed.

2. Understanding the concept of COMMUTATIVE PROPERTIES, which simply means that the numbers (factors) in the problems may be put in a different order, and the answer (product) will still be the same. 2 X 3 = 6 or 3 X 2 = 6

Each day for 10 Days, students will See, Say, Wrap, Rap, Write, Think and Race toward the Mastery of the Multiplication Facts.

Facts are not taught in order. They have purposely been paired so that the least difficult set of facts to learn goes with a set which requires more time and effort.

Daily Lesson Plans (Section 3) give you step by step instructions for each day and the support materials needed.

LEARNING WRAP-UPS® ARE AN INTEGRAL PART OF THIS PROGRAM

How to use them will be discussed on the following pages.

BEFORE YOU BEGIN THE PROGRAM:

Students should each have:

       A folder in which they can put their daily assignments and record charts
       A small bag of manipulatives (macaroni, beans, commerical blocks, etc.)
       A set of Learning Wrap-ups (often referred to as Wrap-ups)
       Colored pencils (preferred) or crayons

The teacher should have:

       A stop watch
       The work pages as needed for each day
       A WRAP-UP RAP audio
       A SKIP COUNTING EXERCISE video
       A SKIP COUNTING EXERCISE audio tape

GETTING READY FOR THE 10 DAYS PROGRAM TO BEGIN

New and improved Learning Wrap-ups come ready to use. A heavy string has already been attached to the top of the set. The set of 10 plastic boards is held together with a rivet for easy classroom management.

Teach students

HOW TO USE THE WRAP-UPS

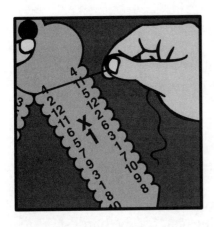 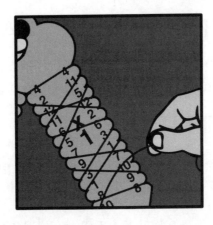 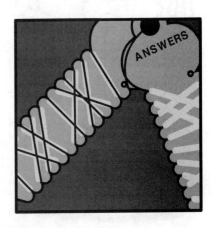

1. Hold the bundle of Wrap-ups at the top of the key and select the specific board you want to wrap.

2. Begin with the string next to the first number on the left. Read the large number in the center and perform the function (example 4x1). Find the answer on the right side, then draw the string around the back to the next number on the left. Continue the process until complete.

3. As you wrap the keys, say the problems and answers aloud. This will provide for auditory learning.

4. When finished, secure the string in the notch at the bottom and see if the string matches the lines on the back. If they do, the key was wrapped correctly.

If you do not have Learning Wrap-ups, check with your local education supply store or call 1-800-992-4966 to find out how you may purchase them.

    LEARNING WRAP-UPS are essential to the success of this program.

PUT WRAP-UPS AWAY by straightening the set, wrapping the string around all 10 boards at once and drawing the string through the notches at the bottom.

10 DAYS TEACHING GUIDE

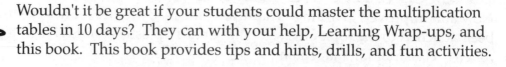

Wouldn't it be great if your students could master the multiplication tables in 10 days? They can with your help, Learning Wrap-ups, and this book. This book provides tips and hints, drills, and fun activities.

By combining the basic teaching concepts of commutative properties of numbers and practice, students can master basic multiplication facts.

THE END GOAL

Students will be able to complete all 10 Wrap-ups in the set in 5 minutes and finish the written test in a specified goal time.

GATHER PRIZES

Gather up small items to be used as prizes (bookmarks, stickers, pencils, key chains, 50¢ gift certificates, etc.). As a teacher, you can ask parents or local merchants to donate items for use during the 10 day program as rewards for those who are quickest, try the hardest, improve the most, reach the teacher-set goal, etc. Do not forget to take advantage of the certificates on pages 123-124 of this book.

TIMING STUDENTS

After several practices, get out your timer or watch the clock. How fast can a Wrap-up be wrapped? Older students should have a goal time of 30 seconds or less. First and second graders should not be timed until they have developed the coordination and confidence to do it in a minute or less. A good way to introduce timing to the young children is to say "Wow you did that fast... would you like me to time you just for fun?" Students will take it from there. See what they are comfortable with.

HOLD FREQUENT WRAP-OFFS

A Wrap-up Wrap-off takes place when 2 or more Wrap-ups are done in one timing. Set a goal:

example:	1 minute for 2 Wrap-ups
later	1 minute for 3 Wrap-ups
	5 minutes for 10 Wrap-ups

FOR MORE WRAP-UP LEARNING

Learning Wrap-ups and teaching guides are available for addition, subtraction, multiplication, division, fractions, pre-algebra, and other subjects.

HELP STUDENTS UNDERSTAND COMMUTATIVE PROPERTY

As students commit each set of math facts to memory, it becomes very clear that they have actually acquired more than the orginal set of facts studied.

The purpose of the following charts is to help the students see clearly that each time they learn the answer to one problem, they have automatically learned the answer to another problem. They will use what we call Mathnique Coloring, a word taken from "technique" and applied specifically to the Learning Wrap-up system of teaching math facts.

SEEING IT IN MATHNIQUE COLOR

Help students to see how quickly they can learn when they approach the times tables with this system.

- Before starting the program, you may choose to give them each a copy of HERE'S WHAT I'VE LEARNED! (p.16), and SEEING IT IN MATHNIQUE COLOR WORKSHEET for the students (p.17). The chart below can be drawn on the chalkboard. As students find the answers, fill in columns 3 and 4.

- Point out that as the problems get harder, there are fewer of them to learn.

- Tell students they will be using the Mathnique Coloring system to help them understand how much they have learned about the times tables.

How many problems are there all together? (144)

1. Color all of the un-colored problems with a factor (number) of:	2. Color them:	3. Number of problems and answers colored.	4. Number of problems re-maining uncolored.
1	YELLOW	23	121
2	YELLOW	21	100
10	ORANGE	19	81
3	ORANGE	17	64
11	RED	15	49
4	RED	13	36
5 & Squares	PURPLE	16	20
6	BLUE	8	12
7	BLUE-GREEN	6	6
9	LIME-GREEN	4	2
8	DARK PINK	2	0
12	You have learned them already!		WOW!!

Point out that after learning about 11, you are over half done.

TEACHER'S CHOICE

1. Read through the entire book, then choose the materials you believe will be needed for your students. Print copies. As specific needs arise, you can provide additional materials that are included in this book.

> *Note:*
> *Teacher's Choice Materials are indicated by TC at the top of the page if it's the whole page, or at the end of the sentence if it's a quick suggestion. This is information to be used at the teacher's discretion.*

2. Determine if you want and need help. There are suggestions, sample letters and other material to copy on pp 19-21.

MANIPULATIVES

Beans, macaroni, buttons, etc. can be used as manipulatives. A zipper type sandwich bag half full of macaroni can be colored by adding two tablespoons of alcohol and a few drops of food coloring. Shake until color is evenly distributed, then spread on newspaper to dry.

MATHNIQUE MONEY TC

Whenever students demonstrate self-control, behavior of an achiever, or are on task, pay them Mathnique Money. It can be used for prize drawings, special privileges, field trips, treats, auctions, white elephant sales, work on a special project, or awards at the end of "10 DAYS". Sample to photocopy on p. 22.

To assure that money does not get lost, have students put their name and date on it and deposit it in the bank which is a sealed box kept on the teachers desk. Students should keep track of their earnings on the cover of their folder. The teacher can check folders periodically to make sure a student will have enough "money" for a particular event or purpose.

Mathnique Money can be given for:
- Attendance
- Returning homework slips
- Returning Learning Wrap-ups
- Reaching goal time, either wrapping or writing
- Helping the teacher
- Helping another student to understand a problem
- Working or starting a class period quietly and without being told to do so
- Being on task
- If one or two students are not on task, quickly pass Mathnique Money to all the others. Let your "Mathnique Money" do the talking.

GETTING HELP

INVOLVE COMMUNITY TC

ASK THEM FOR MERCHANDISE, PRIZES, TICKETS

Many of your students will one day be employees or employers in your community. It is of great importance that the young people we are responsible to teach acquire the skills needed to earn a living. Many of them will start their lifetime of earning in fast food establishments, clerking in stores, selling, etc. We believe it is worthwhile for our community and local merchants to support this most important program.

Send a letter explaining the goals and that you need their help. Invite them to participate in any way they can. Drink coupons, sporting event tickets, video rental, ice cream, candy bars, toys, gift certificates, and pizza coupons are only a few of the items that students enjoy. Ask your Chamber of Commerce, as well as colleges and high schools for promotional products. Sometimes "good as new" items from home can be added to the cache of prizes for students. Some community leaders are willing to speak to your students about math and the importance of it in our lives.

Be sure that your students know that the greatest reward of all will be their knowledge of math facts which they will use every single day of their lives. This knowledge allows them to build towards a greater, more valuable math education.

Letters to merchants could be similar to the one in the next section (p. 19), but phone calls and visits are more successful. Enlist the help of your roommothers and volunteers in this area.

INVOLVE PARENTS TC

ASK THEM FOR HELP AT HOME AND AT SCHOOL

Research has shown that when parents get involved, students have a far greater success rate at school.   Learning Wrap-ups have been very successful with getting parents involved.

Send a letter home explaining your goals and that you need parents to help. Include the STUDENT AGREEMENT (p. 20), some HOMEWORK SLIPS (p. 21), and a TRACK YOUR PROGRESS CHART (p. 18) to be found in the next section.

Be sure that your students know how to use the Learning Wrap-ups. Time students to see how fast they can complete a Wrap-up. Have them record the number of seconds it takes. Always have students take home two Learning Wrap-ups, one Wrap-up for the student and one Wrap-up for the parent.

BEFORE YOU BEGIN, just a few thoughts.....

It has been stated that we remember:

> *20% of what we read, 30% of what we hear,*
> *40% of what we see, 50% of what we say, 60% of what we do*
> ## and **90% of what we see, hear, say and do.** *

Most of us have one sense (visual, auditory or physical) with which we learn more quickly. Research suggests that if we use all 3 of the senses at one time, we can be even more successful.   It is important to help students develop strengths in all three areas and use them whenever possible. This program involves the use of all three senses.

You will find that many accelerated learning basics have been incorporated into the "10 DAY" program.

Remember, however, that you are the teacher.   Judgments will have to be made for each particular child, but always work towards the greatest performance possible from every student. You may be able to speed up the program for your class, or in some cases find it necessary to go slower.

TO HEIGHTEN INTEREST:

> *"We remember best what is at the first and at the last of a session."* *

These suggestions are to be used at the teacher's discretion, but use them frequently. Do something different every day. Keep students guessing and excited.

- Take frequent breaks or change the activity.
- Try CHECKING PROGRESS activities first thing in the morning.
- Workout with the SKIP COUNTING EXERCISE video or audio tape during gym.
- Just before recess have a Wrap-up race.
- Have races with the WebSight Multiplication Writing Sheets. (p. 128)
- Have students sit on the floor and place their writing paper on the chair seat.
- Do Wrap-ups while sitting <u>on</u> their desks.
- Let students sit on the floor back to back.
- Have relays using Wrap-ups. Put them in a sack, have students draw and wrap.
- Play musical chairs. Place a Wrap-up with a different number sticking out on each chair. Students walk around. When the music stops, students wrap and then sit. No one is left without a seat, but students always try to get an easy Wrap-up and don't want to be the last to sit down.
- Have Wrap-up races with the teacher. Invite the principal for a race. Have a parent Wrap-off.
- Group by row, table, hair color, birthday months, number of letters in names, height, color of shoes, velcro or laces, buttons on clothing.   Have students race with Wrap-ups and Write-ons. Regroup often.
- MOST IMPORTANT!! Let students make up some of the activities they think will help them learn. They usually have great ideas. After all, this program is for them. They will give it everything they have if they are part of the planning.

* Accelerated Learning Action Guide by Brian Tracy, with Colin Rose.

SECTION TWO

(Information to photocopy)

HELPER CHART

Keep this paper on your desk. Look at it only if you cannot figure an answer. Do not look until you are sure you need help. As you learn each set of facts, color in the rectangle for the day. Using the same color, fill in the square beside each of the problems and answers you master on that day.   Use a different color every day.

Day 1	Day 2	Day 3	Day 4	Day 5

Day 6	Day 7	Day 8	Day 9	Day 10

☐ 1x1=1	☐ 2x1=2	☐ 3x1=3	☐ 4x1=4	☐ 5x1=5
☐ 1X2=2	☐ 2X2=4	☐ 3X2=6	☐ 4X2=8	☐ 5X2=10
☐ 1X3=3	☐ 2X3=6	☐ 3X3=9	☐ 4X3=12	☐ 5X3=15
☐ 1X4=4	☐ 2X4=8	☐ 3X4=12	☐ 4X4=16	☐ 5X4=20
☐ 1X5=5	☐ 2X5=10	☐ 3X5=15	☐ 4X5=20	☐ 5X5=25
☐ 1X6=6	☐ 2X6=12	☐ 3X6=18	☐ 4X6=24	☐ 5X6=30
☐ 1X7=7	☐ 2X7=14	☐ 3X7=21	☐ 4X7=28	☐ 5X7=35
☐ 1X8=8	☐ 2X8=16	☐ 3X8=24	☐ 4X8=32	☐ 5X8=40
☐ 1X9=9	☐ 2X9=18	☐ 3X9=27	☐ 4X9=36	☐ 5X9=45
☐ 1X10=10	☐ 2X10=20	☐ 3X10=30	☐ 4X10=40	☐ 5X10=50
☐ 1X11=11	☐ 2X11=22	☐ 3X11=33	☐ 4X11=44	☐ 5X11=55
☐ 1X12=12	☐ 2X12=24	☐ 3X12=36	☐ 4X12=48	☐ 5X12=60

☐ 6x1=6	☐ 7x1=7	☐ 8x1=8	☐ 9x1=9	☐ 10x1=10
☐ 6X2=12	☐ 7X2=14	☐ 8X2=16	☐ 9X2=18	☐ 10X2=20
☐ 6X3=18	☐ 7X3=21	☐ 8X3=24	☐ 9X3=27	☐ 10X3=30
☐ 6X4=24	☐ 7X4=28	☐ 8X4=32	☐ 9X4=36	☐ 10X4=40
☐ 6X5=30	☐ 7X5=35	☐ 8X5=40	☐ 9X5=45	☐ 10X5=50
☐ 6X6=36	☐ 7X6=42	☐ 8X6=48	☐ 9X6=54	☐ 10X6=60
☐ 6X7=42	☐ 7X7=49	☐ 8X7=56	☐ 9X7=63	☐ 10X7=70
☐ 6X8=48	☐ 7X8=56	☐ 8X8=64	☐ 9X8=72	☐ 10X8=80
☐ 6X9=54	☐ 7X9=63	☐ 8X9=72	☐ 9X9=81	☐ 10X9=90
☐ 6X10=60	☐ 7X10=70	☐ 8X10=80	☐ 9X10=90	☐ 10X10=100
☐ 6X11=66	☐ 7X11=77	☐ 8X11=88	☐ 9X11=99	☐ 10X11=110
☐ 6X12=72	☐ 7X12=84	☐ 8X12=96	☐ 9X12=108	☐ 10X12=120

HERE'S WHAT I KNOW!

Write all the problems and answers you know. With the
color you have chosen for the day, color the small square beside
each problem and answer you have written. Remember to
do the **COMMUTATIVE PARTNERS** and color them in also.

Day 1	Day 2	Day 3	Day 4	Day 5
▨				

Day 6	Day 7	Day 8	Day 9	Day 10

X 1

- ▨ $1 \times 1 = 1$
- ▨ $2 \times 1 = 2$
- ▨ $3 \times 1 = 3$
- [4]
- [5]
- [6]
- [7]
- [8]
- [9]
- [10]
- [11]
- [12]

X 2

- ▨ $1 \times 2 = 2$
- [2]
- [3]
- [4]
- [5]
- [6]
- [7]
- [8]
- [9]
- [10]
- [11]
- [12]

X 3

- ▨ $1 \times 3 = 3$
- [2]
- [3]
- [4]
- [5]
- [6]
- [7]
- [8]
- [9]
- [10]
- [11]
- [12]

X 4

- [1]
- [2]
- [3]
- [4]
- [5]
- [6]
- [7]
- [8]
- [9]
- [10]
- [11]
- [12]

X 5

- [1]
- [2]
- [3]
- [4]
- [5]
- [6]
- [7]
- [8]
- [9]
- [10]
- [11]
- [12]

X 6

- [1]
- [2]
- [3]
- [4]
- [5]
- [6]
- [7]
- [8]
- [9]
- [10]
- [11]
- [12]

X 7

- [1]
- [2]
- [3]
- [4]
- [5]
- [6]
- [7]
- [8]
- [9]
- [10]
- [11]
- [12]

X 8

- [1]
- [2]
- [3]
- [4]
- [5]
- [6]
- [7]
- [8]
- [9]
- [10]
- [11]
- [12]

X 9

- [1]
- [2]
- [3]
- [4]
- [5]
- [6]
- [7]
- [8]
- [9]
- [10]
- [11]
- [12]

X 10

- [1]
- [2]
- [3]
- [4]
- [5]
- [6]
- [7]
- [8]
- [9]
- [10]
- [11]
- [12]

HERE'S WHAT I'VE LEARNED

Write the answer to each problem <u>down</u> the column, then write the answers for the commutative partners <u>across</u> each row. Choose a different color for each day and lightly color the problems you have mastered. (Mathnique Coloring)

1 x1	1 x2	1 x3	1 x4	1 x5	1 x6	1 x7	1 x8	1 x9	1 x10	1 x11	1 x12
2 x1	2 x2	2 x3	2 x4	2 x5	2 x6	2 x7	2 x8	2 x9	2 x10	2 x11	2 x12
3 x1	3 x2	3 x3	3 x4	3 x5	3 x6	3 x7	3 x8	3 x9	3 x10	3 x11	3 x12
4 x1	4 x2	4 x3	4 x4	4 x5	4 x6	4 x7	4 x8	4 x9	4 x10	4 x11	4 x12
5 x1	5 x2	5 x3	5 x4	5 x5	5 x6	5 x7	5 x8	5 x9	5 x10	5 x11	5 x12
6 x1	6 x2	6 x3	6 x4	6 x5	6 x6	6 x7	6 x8	6 x9	6 x10	6 x11	6 x12
7 x1	7 x2	7 x3	7 x4	7 x5	7 x6	7 x7	7 x8	7 x9	7 x10	7 x11	7 x12
8 x1	8 x2	8 x3	8 x4	8 x5	8 x6	8 x7	8 x8	8 x9	8 x10	8 x11	8 x12
9 x1	9 x2	9 x3	9 x4	9 x5	9 x6	9 x7	9 x8	9 x9	9 x10	9 x11	9 x12
10 x1	10 x2	10 x3	10 x4	10 x5	10 x6	10 x7	10 x8	10 x9	10 x10	10 x11	10 x12
11 x1	11 x2	11 x3	11 x4	11 x5	11 x6	11 x7	11 x8	11 x9	11 x10	11 x11	11 x12
12 x1	12 x2	12 x3	12 x4	12 x5	12 x6	12 x7	12 x8	12 x9	12 x10	12 x11	12 x12

SEEING IT IN MATHNIQUE COLOR

You will need a copy of HERE'S WHAT I'VE LEARNED! (p.16) to complete this worksheet.

Follow the directions for columns 1, 2, 3, and 4 in order before going on to the next number in column 1.

Then think about and answer the questions below.

1. Color all of the un-colored problems with a factor (number) of:	2. Color them.	3. Number of problems and answers colored.	4. Number of problems remaining uncolored.
1	YELLOW		
2	YELLOW		
10	ORANGE		
3	ORANGE		
11	RED		
4	RED		
5 & Squares	PURPLE		
6	BLUE		
7	BLUE-GREEN		
9	LIME-GREEN		
8	DARK PINK		
12	You have learned them	already!	WOW!

1. How many problems are there on page 16 to start? _____

2. The numbers in column 4 are the answers to all of the problems in which a number is multiplied by itself. They are called "perfect squares". Use a black crayon to outline the squares which have perfect square problems.

3. How many "perfect square" problems are there?_____

4. Suppose you could learn 2 sets of multiplication facts a day, write down the order in which you would choose to learn them. _____

5. How many days would it take? _____

6. Could you learn more than 2 sets a day? _____ Which ones? _____

7. If you wanted to learn more than two sets a day, how would you group them?
_____

8. Explain why you would do it the way you have chosen. Consider the end, when there are only 5 or less facts to learn for each set of factors (problems).
_____
_____

17

NAME _____ # _____

TRACK YOUR PROGRESS

Start at the bottom and write the seconds it takes to do a Wrap-up above the number of the Wrap-up you are using. Do it over and over. The faster you wrap the higher you climb, and the fewer seconds it takes. When you meet one of the goals on the chart below, mark it also.

A favorite slogan......

Good, Better, Best,
Never let it rest,
Till the Good gets Better
and the Better gets Best!

—Anonymous

Set goals as suggested on the chart below. Put an X in the square when you meet a particular goal. Just keep wrapping! The faster you get, the better you know it.

Reaching Goals

10 sec.										Next to impossible!
15 sec.										Incredible!
20 sec.										Super Champ!
30 sec.										Champion!
45 sec.										First Goal Keep it up!
1 min.										Good work! You're Learning!
2 min.										You're getting acquainted
1	2	3	4	5	6	7	8	9 10		

Keeping Track

1	*2*	*3*	*4*	*5*	*6*	*7*	*8*	*9*	*10*

18

Dear

In two weeks our class will start an intense 10 DAY program to master the multiplication facts. We know it can put us a "jump ahead" in our math program. "10 DAYS TO MULTIPLICATION MASTERY" has been very successful in other schools, and we expect our students to accomplish the same goals. It is essential that students be in attendance every day, so we are asking for help in motivating them.

Will you please be part of the program by providing some small prizes that students can earn by:

         1. Being in attendance every day

         2. Making the most progress

         3. Practicing at home the most total time

         4. Successfully reaching the goal

Thank you for your support. We will encourage our students and their parents to support your business also.

If it is convenient, may I stop by on_____.

Thank you again.

Teacher
Grade
School
Phone

Dear Parent,   date

Next week our class will start an intense 10 DAY program to master the multiplication facts. We need your help and support during these next two weeks. **It is essential that your student be in attendance every day.** We will be working with fun and effective teaching tools called Learning Wrap-ups, which your student will be bringing home. Please have your child show you how to use them, then have races with each other. Time how long it takes to finish a Wrap-up. Encourage your student to do the very best possible.

Please record the amount of time spent practicing and working on the multiplication facts at home each evening and return the slip along with the WRAP-UPS the following morning.

Students will be recognized for the amount of time they have practiced, how fast they can do a Learning Wrap-up, and how high they score on a written test. There is a direct correlation between time spent practicing at home and success on the tests. It is even better when you get involved.

If possible, could you spend some time in the classroom with us? We need timers, correctors, and parents who will encourage, cheer, and applaud.

Thanks for your cooperation during this most important 10 DAYS.

Sincerely,   Starting Date

Please check if you can help us and ask your student to sign the agreement below and return it to school.
Cut only the top dotted line.   ✂

☐ Yes, I can help! ☐ Maybe I can help. Call me at   Parent's Signature

---

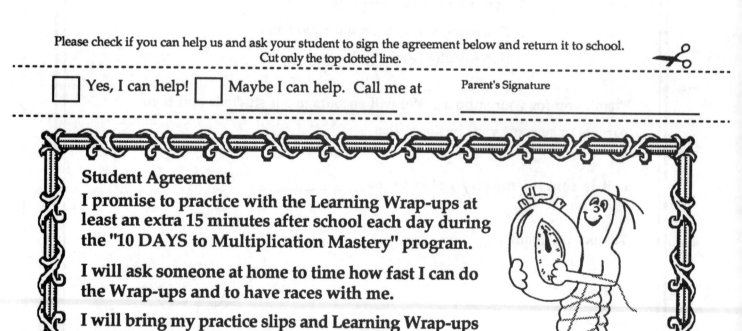

Student Agreement

I promise to practice with the Learning Wrap-ups at least an extra 15 minutes after school each day during the "10 DAYS to Multiplication Mastery" program.

I will ask someone at home to time how fast I can do the Wrap-ups and to have races with me.

I will bring my practice slips and Learning Wrap-ups back to school each day so my name can be put in the drawing box. If my name is drawn, I will win a prize.

Student's Signature

HOMEWORK SLIPS

Please clip and return a signed student practice slip to school
each day. Practice at home is reflected in the progress at school.
Please help your student to have quality practice time.
Thank you.

Student's Name   #

Practiced Math for _____ minutes. _____
                                              date

Parent Signature

Student's Name   #

Practiced Math for _____ minutes. _____
                                              date

Parent Signature

Student's Name   #

Practiced Math for _____ minutes. _____
                                              date

Parent Signature

Student's Name   #

Practiced Math for _____ minutes. _____
                                              date

Parent Signature

Student's Name   #

Practiced Math for _____ minutes. _____
                                              date

Parent Signature

Student's Name   #

Practiced Math for _____ minutes. _____
                                              date

Parent Signature

Student's Name   #

Practiced Math for _____ minutes. _____
                                              date

Parent Signature

Student's Name   #

Practiced Math for _____ minutes. _____
                                              date

Parent Signature

Student's Name   #

Practiced Math for _____ minutes. _____
                                              date

Parent Signature

Student's Name   #

Practiced Math for _____ minutes. _____
                                              date

Parent Signature

MATHNIQUE ™
Money
You are
ACHIEVING!

Write name and deposit in the MATHNIQUE Bank

MATHNIQUE ™
Money
You are
ACHIEVING!

Write name and deposit in the MATHNIQUE Bank

MATHNIQUE ™
Money
You are
ACHIEVING!

Write name and deposit in the MATHNIQUE Bank

MATHNIQUE ™
Money
You are
ACHIEVING!

Write name and deposit in the MATHNIQUE Bank

MATHNIQUE ™
Money
You are
ACHIEVING!

Write name and deposit in the MATHNIQUE Bank

MATHNIQUE ™
Money
You are
ACHIEVING!

Write name and deposit in the MATHNIQUE Bank

MATHNIQUE ™
Money
You are
ACHIEVING!

Write name and deposit in the MATHNIQUE Bank

MATHNIQUE ™
Money
You are
ACHIEVING!

Write name and deposit in the MATHNIQUE Bank

MATHNIQUE ™
Money
You are
ACHIEVING!

Write name and deposit in the MATHNIQUE Bank

MATHNIQUE ™
Money
You are
ACHIEVING!

Write name and deposit in the MATHNIQUE Bank

SECTION THREE
Let's Go!

LESSON PLANS AND WORKSHEETS FOR EACH DAY

DAY 1

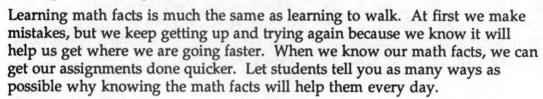

- Pass out folders, charts and worksheets for today. Students should keep these at their desks. Tell students to keep all their work in the folder as you will be checking it periodically.

TALK ABOUT

- Talk to the students about when they first learned to walk. Tell them about the hundreds of times they fell down, then got up and tried again. Ask them why they kept getting up and trying again. (Possible answer: They could get from one place to another faster.)

  Learning math facts is much the same as learning to walk. At first we make mistakes, but we keep getting up and trying again because we know it will help us get where we are going faster. When we know our math facts, we can get our assignments done quicker. Let students tell you as many ways as possible why knowing the math facts will help them every day.

- Talk about the properties of 1. Explain that any number multiplied by 1 is that number. Stack 3 books and explain that you have only 1 stack. How many books are there? WRITE $1 \times 3 = 3$ on the board. Place the 3 books on the table separately and write $3 \times 1 = 3$. Show that the answer is the same.

TODAY'S LESSON #1 AND #2

- Use the CIRCLE PATTERN (p. 27) and a small bag of manipulatives to work several problems, 1X3, 3X1; 8X1, 1X8; etc. Work the problems on the circle pattern sheet. Point out that regardless of the arrangement, the answer is still the same for each pair of problems. Write each of the problems so the students will become familiar with what the multiplication problem looks like. Show both vertical and horizontal problems.

- Explain commutative property. Webster's New World Dictionary says: "Commutative: 1. involving exchange or replacement 2. of or pertaining to an operation in which the order of the elements does not affect the result, as, in addition, $3 + 2 = 2 + 3$ and, in multiplication $2 \times 3 = 3 \times 2$."

  EXAMPLES: If Brent and Abby were riding the bus to school, it wouldn't matter if Brent rode in the front and Abby rode in the back, or if Abby rode in the front and Brent rode in the back. They would still get to school.

  What if all the boys sat in the front and the girls sat in the back? Or maybe all the girls sat in the front and the boys sat in the back? Would they still get to school?

  With multiplication problems, the same is true. Two factors (numbers in the problem) can be reversed and still get the same answers.

Day One (cont)

- Pass out the Wrap-ups. Students separate the #1 board for wrapping and hold the other boards all together.

- Show the answer lines on the back of the #1 board. Explain that the lines on the back are for the students to correct their own work. The teacher never looks at the back.

- Tell your students that Wrap-ups will help them learn the facts fast, and as the teacher you will know they have been doing it right when they do the RAPID WRITING EXERCISE WORKSHEET tomorrow.

> *It is wise to walk about the classroom to determine if the students are using the Wrap-up correctly. If they are not, you can whisper the correct way to do them, or even help by holding their hands and doing a few problems together.*

- Practice with the #1 Wrap-ups for about ten minutes or until they are comfortable with it.

> *Have students say the problems and answers aloud until they can do the whole board in 40 seconds. This is an important part of getting the information into long term memory. Then continue practice for speed.*

- Have students write the answers on the CONCENTRATION DRILL STRIP WORKSHEET (p. 28) so they can see the commutative concept.

> *We're told that a picture is worth a thousand words. This is true for our memories. The pictures that the students draw on the COMMUTATIVE WORKSHEETS, and the visual memory of the commutative partners they write at the bottom, stimulate their mind to rearrange the problem in order to come up with an answer when necessary.*

- Do the COMMUTATIVE X1 WORKSHEET. (p. 29) Read the story problems to the students and work them together. Check students' papers and drawings as you walk about the classroom. It is essential that they get the format and problems right. If they make errors, whisper to them so errors can be corrected **immediately**. The time you spend today making sure everyone understands will pay off with success in future assignments.

> *Make certain your volunteers know the importance of moving about the classroom and correcting students with whispers.*

Day One (cont)

- SKIP COUNT the 2's with the SKIP COUNTING AUDIO TAPE five to ten times. If students can already count by 2, move to the next step.

- Get out the multiplication Wrap-ups. Do the #2 Wrap-up together, having the students say the problems and answers aloud. Practice as long as possible.

> *Vince Lombardi once said:*
>   *"Practice does not make perfect. PERFECT PRACTICE MAKES PERFECT."*
>
> Learning Wrap-ups provide PERFECT PRACTICE! Students see it, say it, hear it, and do it. They know instantly if they have made an error, and in seconds can be doing it perfectly correct.

- Do the COMMUTATIVE X2 WORKSHEET (p. 30)

- During Physical Education, exercise by 2's to the SKIP COUNTING VIDEO.

- <u>Other X2 material to use if needed:</u>
      DRILL STRIP WORKSHEET (p. 31)
      INDIVIDUALIZED STORY PROBLEMS WORKSHEET (p. 32)
      X2 WRAP-UP RAP Audio
      CIRCLE PATTERN AND MANIPULATIVES for X2 (p. 27)

SEND HOME

- Learning Wrap-ups and the Parent Packet (which includes: Letter to the Parents, Student Agreement, Track Your Progress Chart, and Homework Slips).
  Practice the #1 and #2 boards.
  GOAL: To do each Wrap-up in 30 seconds or less and record it on the TRACK YOUR PROGRESS CHART at home.

> *Parents generally get intensely involved with their students' learning while using Wrap-ups together.*

- Inform students they must bring Learning Wrap-ups back the next day. (Students who forget their Wrap-ups must cut up the CONCENTRATION DRILL STRIPS X2 WORKSHEET (p. 31) and use it for practice while other students are using Wrap-ups.)

- Remind them about prize drawings for bringing back their signed homework slips and Learning Wrap-ups. TC

CIRCLES PATTERN
KEEP PATTERN IN FOLDER TO USE AGAIN.

Arrange manipulatives to show a variety of multiplication problems. Examples: $2 \times 6, 6 \times 2$
$1 \times 10, 10 \times 1$ $7 \times 3, 3 \times 7$ $5 \times 4, 4 \times 5$ etc.

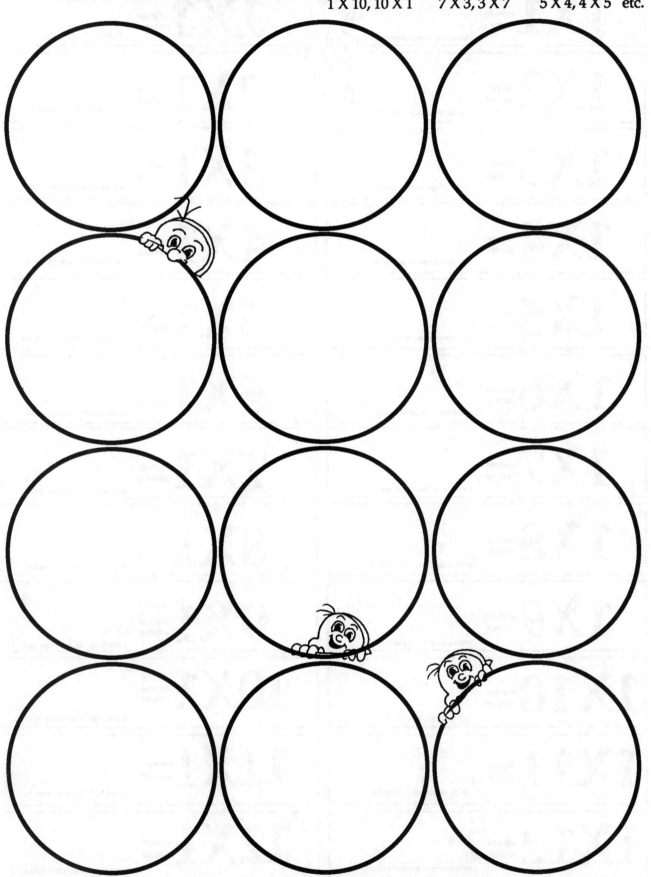

1X1= _____	1X1= _____
1X2= _____	2X1= _____
1X3= _____	3X1= _____
1X4= _____	4X1= _____
1X5= _____	5X1= _____
1X6= _____	6X1= _____
1X7= _____	7X1= _____
1X8= _____	8X1= _____
1X9= _____	9X1= _____
1X10= _____	10X1= _____
1X11= _____	11X1= _____
1X12= _____	12X1= _____

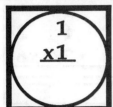

1 x1	2 x1	1 x2	3 x1	1 x3	4 x1	1 x4	5 x1	1 x5	6 x1	1 x6

1x7= ___ 7x1= ___	1x8= ___ 8x1= ___	1x9= ___ 9x1= ___	1x10= ___ 10x1= ___	1x11= ___ 11x1= ___	1x12= ___ 12x1= ___

One of the sacks has been drawn for you!

1. Justin had 6 sacks. Each sack had 1 orange. Draw a picture of his sacks and oranges. Write a multiplication problem and answer to show how many oranges Justin had.

_____

2. Kathy had 1 box with 6 oranges in it. Draw a picture of the oranges in the box. Write a multiplication problem and answer to show how many oranges Kathy had.

Here is the box.

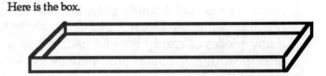

_____

3. Josh had 8 circles with 1 macaroni in each circle. Draw a picture of his circles and macaroni. Write a multiplication problem and answer to show how many pieces of macaroni Josh had in all.

The first circle and macaroni are already done.

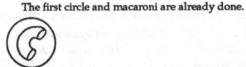

_____

4. Jenna used only 1 circle. She put 8 pieces of macaroni in that circle. Draw the macaroni in the circle. Write a multiplication problem and answer to show how many pieces of macaroni Jenna had in all.

Here is the circle!

_____

5. Write all of the 1X ___ = ___ problems and their commutative partners ___ X 1 = ___ .

2 x1	1 x2	2 x2	3 x2	2 x3	4 x2	2 x4	5 x2	2 x5	6 x2	2 x6

7x2 = ___ 2x7 =___	2x8 = ___ 8x2 =___	2x9 = ___ 9x2 = ___	2x10 = ___ 10x2 = ___	2x11 = ___ 11x2 = ___	2x12 = ___ 12x2 = ___

1. Chris had 5 candy bars. Each bar had 2 sections. Draw a picture to show how many sections of candy bars Chris had. Write a multiplication problem and answer to show how many sections he had in all.

The first candy bar is done for you!

_____

2. Andrea only had 2 candy bars, but she cut each one into 5 pieces. Draw a picture to show how many pieces of candy bar she had. Write a multiplication problem and answer to show the total pieces.

The first candy bar is done for you!

_____

3. There were 7 houses with 2-car garages on Greg's street. Draw a picture of the garages and cars. Write a multiplication problem and answer to show how many cars there would be if all the garages were full.

The first garage is done for you!

_____

4. Shirley's brother had 2 big garages in his toy car set. Each garage holds 7 toy cars. Draw a picture of the garages and cars. Write a multiplication problem and answer that shows how many cars there were in all.

The first toy garage is done for you!

_____

5. Write all of the 2X ___ = ___ problems and their commutative partners   ___ X 2 =____ .

_____   _____   _____   _____

_____   _____   _____   _____

_____   _____   _____   _____

_____   _____   _____   _____

_____   _____   _____   _____

30

$2 \times 1 =$ _____	$1 \times 2 =$ _____
$2 \times 2 =$ _____	$2 \times 2 =$ _____
$2 \times 3 =$ _____	$3 \times 2 =$ _____
$2 \times 4 =$ _____	$4 \times 2 =$ _____
$2 \times 5 =$ _____	$5 \times 2 =$ _____
$2 \times 6 =$ _____	$6 \times 2 =$ _____
$2 \times 7 =$ _____	$7 \times 2 =$ _____
$2 \times 8 =$ _____	$8 \times 2 =$ _____
$2 \times 9 =$ _____	$9 \times 2 =$ _____
$2 \times 10 =$ _____	$10 \times 2 =$ _____
$2 \times 11 =$ _____	$11 \times 2 =$ _____
$2 \times 12 =$ _____	$12 \times 2 =$ _____

PICKING UP SEASHELLS x2

Two students, on their way to school, found a big can filled with seashells.
It was accidently dropped in our classroom and the shells spilled all over.
Fill in the spaces with names of students in our classroom.

	Answers	Write the problems and show how you were thinking to get the answers.

1. _____ and _____ each picked up 8 matching shells. How many matching shells were picked up by the 2 students?

2. _____ picked up only the shells that were 2 inches wide. He found 7 of them. If the shells were laid side by side, how many inches were there all together?

3. _____ picked up 6 flat white shells about the size of a mustard jar lid. _____ picked up the same number of shells that looked like a porcupine. How many shells did the 2 students pick up ?

4. _____ said he would pay 5¢ each for 2 little pure white shells. How much would that be?

5. _____ looked and looked for shells with pointed ends. She put brown pointed shells in one pile and white pointed shells in another. When she had found them all, she counted 9 shells in each of the 2 piles. How many pointed shells were there?

6. _____found smooth black rocks that looked almost like shells. He held 4 rocks in each hand. How many rocks was he holding?

7. _____, _____, and _____ finally put all the shells back in the can. The 3 students each put 2 hands full in the can. How many handfuls of shells were there?

DAY 2

CHECKING PROGRESS

- Collect HOMEWORK SLIPS. Read aloud the amount of time some students spent practicing. (You do not need to read their names. They know who they are.) PRAISE! Reward!

- Pass out TRACK YOUR PROGRESS CHARTS to be kept in the folders. (p. 18 section 2)

- Wrap #2 Learning Wrap-up as fast as they can. Have students record their completion time on the chart.

- Have students complete the RAPID WRITING EXERCISE (p.35). Explain that it is important to learn to write fast. Count the seconds softly so that students can record how fast they do each column.

- Pass out HELPER CHART found on p. 14.
    Color in rectangle above Day 1. Teacher chooses the color. (see this book cover)
    Color all the rectangles beside column 1.
    Color all the rectangles beside 1X __ at the top of each column.
    Color all the rectangles beside column 2.
    Color all the rectangles beside 2X__ in each column.

TALK ABOUT

- Discuss the commutative property of numbers in a multiplication problem. Help students discover that by learning 24 facts (all the 1's and all the 2's) they actually know 44 facts. Have them count the colored rectangles on their HELPER CHART. Praise them for learning 44 facts in one day.

  Have students keep the HELPER CHART in their folders for reference any time they need to know an answer or to color in the new facts as they are learned.

TODAY'S LESSON #10 and #3

- Talk about the properties of 10. Compare the properties of 10's and 1's.

- Do the DRILL STRIP X10 WORKSHEET.

- Do COMMUTATIVE X10 WORKSHEET. Read the story problems to the students and then work the problems as a class.   Talk to them about the thought process they use to solve the problems. Tell them it is important to be able to put their thoughts into words. Invite them to share how they solved a particular problem. Whatever they tell you is OK.

- Work with the #10 Learning Wrap-up. Practice 10 minutes. Have students say problems and answers aloud until they can do it in 40 seconds. When 10 minutes are up, have them do it once more while you softly count the seconds. Have them record their completion time on their TRACK YOUR PROGRESS CHART.

Day 2 (cont)

- Work with the #3 Learning Wrap-up. On the board, write the problems and answers in the same order as on the Wrap-up. Have students do Wrap-ups together while saying problems and answers aloud. Erase one problem each time they do the Wrap-up, starting with 1x3, 2x3, etc. Continue until all the problems are erased.

- Do the CONCENTRATION x3 DRILL STRIP WORKSHEETS. If an student leaves their Learning Wrap-ups at home they will need to cut up the drill strips and use them for flash card practice at school.

- **If manipulatives are necessary, use the circles pattern for some 3X__ and __X3 problems.**

- **Do the COMMUTATIVE X3 WORKSHEET. Equate this work with getting up and trying to walk again and again.**

- Work with the X3 Learning Wrap-ups. Say problems and answers aloud until they can do the Wrap-up in 40 seconds.

- **Do INDIVIDUALIZED STORY PROBLEMS "AT SCHOOL"**

- Other materials available for the 3's are:

    CIRCLES PATTERN for work with manipulatives if necessary
    NUMBER FAMILY X3 FUNSHEET
    SKIP COUNTING BY 3's video exercises
    SKIP COUNTING BY 3's audio with music
    WRAP-UP RAP audio to do separately or to go with the Learning Wrap-ups

---

Get Students involved. Have some fun with the Individualized Story Problems. Read them aloud, then have students put in the names of someone in the class. Extend practice by counting the letters in each person's name and have the students multiply by 3. Have them write one story problem of their own on a separate piece of paper and turn it in. You could compile them for a story problem workbook later on.

Dancing or moving about while reciting the problems and answers with the Wrap-up Rap is a favorite of the students, and works great for a break in the middle of the class period.

A WELL-KNOWN SECRET: If you are walking about the classroom as students do their worksheets and Learning Wrap-ups, you will observe if students are doing their work correctly. If they are not, you can whisper in their ear so it can be changed. It will not be necessary to collect and correct every paper. Have students put worksheets in their folders. You can check the folders of the students with whom you are concerned.

SEND HOME

- Learning Wrap-ups. GOAL: To do the #3 board as fast as they can do the #10 Learning Wrap-up. Work towards wrapping the boards in 30 seconds.

- **Tell students to race their parents.**

- Remind students to bring back their Learning Wrap-ups and their signed homework slips. Tell them that when they practice at home, you can tell by how well they do on the Wrap-up and writing exercises the next day. Suggest that the teacher knows when the amount of time they say they practiced does not coincide with the results the next day.

- Praise them for the work they have accomplished.

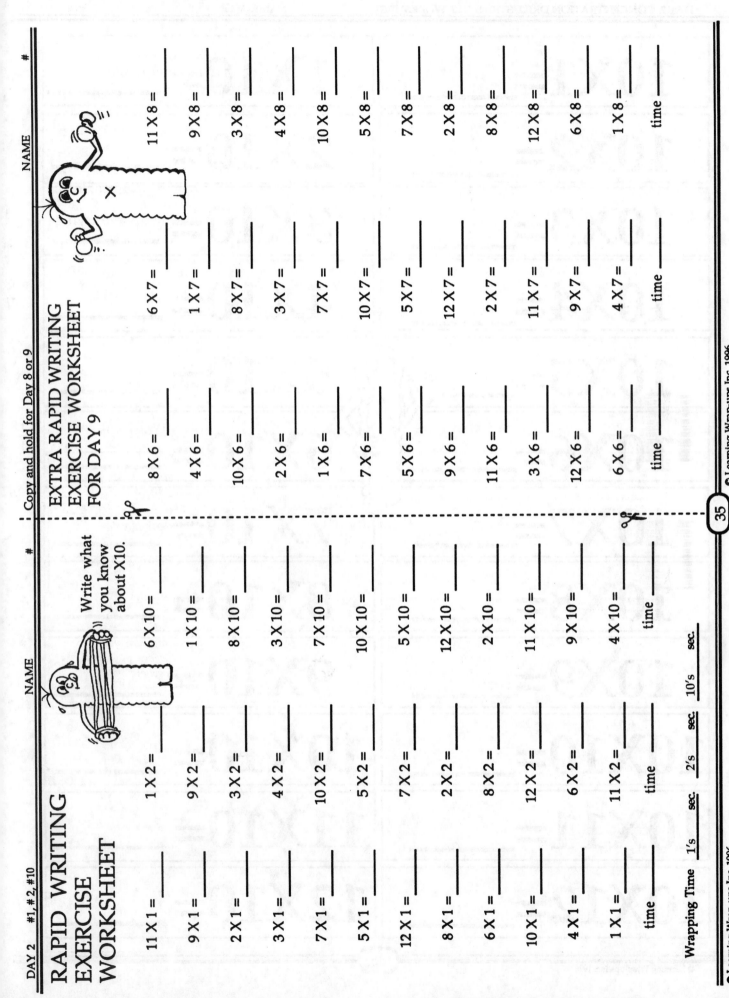

DAY 2   #1, #2, #10

RAPID WRITING EXERCISE WORKSHEET

NAME _____   # _____

Write what you know about X10.

11 X 1 =
9 X 1 =
2 X 1 =
3 X 1 =
7 X 1 =
5 X 1 =
12 X 1 =
8 X 1 =
6 X 1 =
10 X 1 =
4 X 1 =
1 X 1 =
time _____

1 X 2 =
9 X 2 =
3 X 2 =
4 X 2 =
10 X 2 =
5 X 2 =
7 X 2 =
2 X 2 =
8 X 2 =
12 X 2 =
6 X 2 =
11 X 2 =
time _____

6 X 10 =
1 X 10 =
8 X 10 =
3 X 10 =
7 X 10 =
10 X 10 =
5 X 10 =
12 X 10 =
2 X 10 =
11 X 10 =
9 X 10 =
4 X 10 =
time _____

Wrapping Time   1's _____ sec.   2's _____ sec.   10's _____ sec.

© Learning Wrap-ups Inc. 1996

Copy and hold for Day 8 or 9

EXTRA RAPID WRITING EXERCISE WORKSHEET FOR DAY 9

8 X 6 =
4 X 6 =
10 X 6 =
2 X 6 =
1 X 6 =
7 X 6 =
5 X 6 =
9 X 6 =
11 X 6 =
3 X 6 =
12 X 6 =
6 X 6 =
time _____

6 X 7 =
1 X 7 =
8 X 7 =
3 X 7 =
7 X 7 =
10 X 7 =
5 X 7 =
12 X 7 =
2 X 7 =
11 X 7 =
9 X 7 =
4 X 7 =
time _____

NAME _____   # _____

11 X 8 =
9 X 8 =
3 X 8 =
4 X 8 =
10 X 8 =
5 X 8 =
7 X 8 =
2 X 8 =
8 X 8 =
12 X 8 =
6 X 8 =
1 X 8 =
time _____

35

© Learning Wrap-ups Inc. 1996

10X1=____	1X10= ____
10X2=____	2X10= ____
10X3=____	3X10= ____
10X4=____	4X10= ____
10X5=____	5X10= ____
10X6=____	6X10= ____
10X7=____	7X10= ____
10X8=____	8X10= ____
10X9=____	9X10= ____
10X10=____	10X10= ____
10X11=____	11X10= ____
10X12=____	12X10= ____

| 1 10 x10 x1 | 2 10 x10 x2 | 3 10 x10 x3 | 4 10 x10 x4 | 5 10 x10 x5 | 10x6=___ 6x10=___ |

| 10x7=___ 7x10=___ | 10x8=___ 8x10=___ | 10x9=___ 9x10=___ | 10 x10 | 10x11= ___ 11x10= ___ | 10x12= ___ 12x10= ___ |

1. Ammon put his candy in rows with 10 in each row. He had 3 rows. Draw a picture to show how Ammon's candy looked. Write the multiplication problem and answer that shows how much candy he has.

   _____

The first row is done for you!

The first column is done for you!

2. Michelle has 3 little marshmallows in each column. She has 10 columns. Draw a picture to show how Michelle's marshmallows look. Write the multiplication problem and answer that shows how many marshmallows Michelle has.

   _____

The top raisin in each column is done for you!

3. Beth wanted to have the exact number of raisins in each cookie. So she lined up the raisins. She made 10 columns with 5 raisins in each column. Draw a picture of all the raisins lined up. Write a problem and answer to show how many raisins Beth had.

   _____

The first cookie is done for you!

4. Mo has 5 cookies. Each cookie has 10 raisins. Draw a picture showing her cookies and raisins. Write a multiplication problem to show how many raisins are in the 5 cookies.

   _____

5. Write all of the 10X __ = __ problems and their commutative partners___ X10 =___.

3X1= _____	1X3= _____
3X2= _____	2X3= _____
3X3= _____	3X3= _____
3X4= _____	4X3= _____
3X5= _____	5X3= _____
3X6= _____	6X3= _____
3X7= _____	7X3= _____
3X8= _____	8X3= _____
3X9= _____	9X3= _____
3X10= _____	10X3= _____
3X11= _____	11X3= _____
3X12= _____	12X3= _____

3 1 x1 x3	3 2 x2 x3	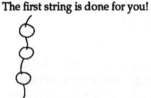 3 x3	4 3 x3 x4	5 3 x3 x5	6 3 x3 x6
7x3= ___ 3x7=___	3x8= ___ 8x3=___	3x9= ___ 9x3= ___	3x10= ___ 10x3= ___	3x11=___ 11x3= ___	3x12= ___ 12x3= ___

1. Caryn had 8 beads on each string. She had 3 strings. Draw a picture showing Caryn's beads and strings. Write a multiplication problem and answer to show how many beads Caryn had.

The first string is done for you!

_____

2. Rob had 8 strings, but he only had 3 beads on each string. Draw a picture showing Rob's beads and strings. Write a problem and answer to show how many beads Rob had in all.

The first string is done for you!

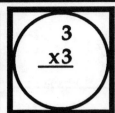

_____

3. Natalie's game took 7 marbles for each player. There were 3 players. Draw the marbles by each player. Write the problem and answer that shows how many marbles are needed for everyone to play.

The first person's marbles are done for you!

_____

4. Mark's game needs only 3 marbles per player. He put each player's marbles in a little cup. He has 7 cups. Draw the marbles in the cups. Write a multiplication problem and answer that tells how many marbles Mark has.

The first cup is done for you!

_____

5. Write all of the 3X ___ = ___ problems and their commutative partners   ___ X 3 =____.

_____   _____

_____   _____

_____   _____

_____   _____

AT SCHOOL Multiply by 3

	Answer	Show how you were thinking to find the answer.

1. The principal, _____, asked 3 students each from 5 classes to be on the program at PTA. How many children were on the program?

2. _____ made 6 piles of counters. Each pile had 3 counters in it. How many counters were there?

3. _____ had 3 pairs of shoes. A pair is 2. How many shoes does she have?

4. When _____ got home from school he made some snowballs. He had 3 stacks and put 8 snowballs in each stack. How many snowballs did he have all together?

5. _____ counted 9 chairs on each row. There were 3 rows. How many chairs did she count?

6. _____ told the students about 3 sets of twins that he knew. How many twins were there?

7. _____ put 4 books on each shelf. There were 3 shelves. How many books did she put away?

8. _____ noticed that 5 children had shirts with 3 buttons on them. How many buttons would that be all together?

9. Our teacher, _____, gave gumdrops to the students who finished their work. 9 students earned 3 gumdrops each. How many gumdrops were earned?

NUMBER FAMILY FUNSHEET
X3 or 3X
SKIP COUNTING BY 3

Color each product and its two sets of factors the same color. Use a different color for each product.

Write "the skip counting by 3" numbers in the TRIANGLES. Do it three times.

3
6
9
12
15
18
21
24
27
30
33
36

Tree numbers: 36, 7, 13, 6, 26, 8, 16, 12, 2, 1, 24, 22, 9, 3, 15, 31, 21, 5, 27, 33, 18, 30, 23

Color all the products in the X3 number family green.

Factors	Product	Factors
3X1	3	3X3
3X4	6	1X3
3X5	9	4X3
3X2	12	2X3
3X3	15	5X3
3X6	18	7X3
3X8	21	6X3
3X10	24	8X3
3X7	27	12X3
3X12	30	9X3
3X9	33	11X3
3X11	36	10X3

Shade all of the numbers that are **not** part of the X3 number family.

1	2	3	4	5	6	7	8	9	10
11	12	13	14	15	16	17	18	19	20
21	22	23	24	25	26	27	28	29	30
31	32	33	34	35	36	37	38	39	40

DAY 3

CHECKING PROGRESS

- Collect parent signed homework slips. Draw for prize(s). Tell students if they practiced, they will do their Wrap-ups faster and do well on their RAPID WRITING EXERCISE. Let them know you will be comparing the homework time with other activities.

- Do the #10 and #3 Learning Wrap-ups. Time and record on the TRACK YOUR PROGRESS CHART. (in folder)

- Do Day 3 RAPID WRITING EXERCISE. Explain again that it is important for students to learn to write fast. Ask students to write their Wrap-up time for the #3 and #10 boards at the bottom of the sheet.

- Collect #10 Learning Wrap-up and the RAPID WRITING EXERCISE. Praise!

NOTE: The RAPID WRITING EXERCISE is designed for the teacher to assess the students' knowledge. Quickly check the particular students with whom you may be concerned. If a student has not passed off all four facts (1, 2, 3, 10) be sure the Wrap-ups are sent home. Call the parent and ask for support.

- Fill in the HELPER CHART (in folder) both down and across for #3 and #10.   Note the commutative learning process!!! Talk about it again and again.

- Stress importance of attendance! Give treats!

> *TREATS simply tell students that YOU KNOW they are doing something right, that they are succeeding at a particular skill or are trying hard. They do not need to be much, cinnamon candy, minature marshmallows, mathnique money (p. 22) etc.   Students should also understand that you give treats because YOU want to. Asking or telling you they deserve a treat should eliminate the possibility of getting one. If businesses are supporting your program, be sure to tell students who provided the treat and indicate appreciation. Suggest they stop by the business and thank the manager.*

TODAY'S LESSON #11 and #4

There is not an individual Wrap-up for #11; however, 11X problems are on each of the Wrap-ups. They are so much fun that it would be a shame to miss 11's. Take a few minutes to let the students discover some information about the 11 times tables.

- Do the MULTIPLYING BY ELEVEN WORKSHEET (p. 45). Older students could simply count the spaces and write in every eleventh number.

- Do #11 RAPID WRITING EXERCISE (p. 46). Allow no more than 45 seconds per column. Students should draw a line under the last problem answered in the time limit, then finish the column.

- Talk to students about discovering something wonderful. Tell them they will be able to amaze their parents with this next math trick.

Day 3 (cont)

- Introduce JUST FOR FUN, AN EXTENDED ACTIVITY (p. 46). Read through the directions and do a few problems on the board. Let students work the problems on the page.

- Have students write the answers on the CONCENTRATION X 4 DRILL STRIP (p.47) WORKSHEETS. Talk to the students about what they may have noticed concerning the answers they have written. (They all end in even numbers.) Ask students, "When I multiply by 4, can I ever get an odd number in the ones position?" Did they notice that any number multiplied by 4 is double that same number multiplied by 2?

- Play and count aloud with the SKIP COUNT by 4's tape five times.

- Do the COMMUTATIVE X 4 WORKSHEET (p.48). Help students with the story problems. Do them together as a whole class. Continue to check their drawings.

- Do the NUMBER FAMILY X4 FUNSHEET (p.49)

- Do the WATCHING T.V. X4 WORKSHEET (p.50)
  Have students write in the names of their family (brothers, sisters, parents, grandparents, cousins). Talk to students about writing down their thought processes to solve the problems. If they have not done this before, ask questions to help them.
    Did you make marks and count them? Did you multiply by 2 and double it?
    Did you skip count up to the answer? Did you count by 5 and subtract a number?
    Did you make columns and rows of things? Every answer is OK. Congratulate the
        students who were willing to talk about how they solved the problem. Suggest
        that all the students think about it, and write it down in the space allowed.

- Work with the #4 Wrap-ups.

  Other materials available for X4:

        ARRAY PATTERN for manipulatives. (Keep in folder for quick lessons.)
        SKIP COUNTING EXERCISE X4 AUDIO (Make up their own exercises.)
        WRAP-UP RAP X4 AUDIO

SEND HOME

- Just before going home, have students skip count by 4, and work four or five numbers multiplied by 11.

- Learning Wrap-ups. Practice boards #3 and #4. GOAL: Do each board in 40 seconds or less. Work towards 30 seconds. Tell students to astound their parents with their new knowledge of multiplying by 11.

- Remind students about returning homework slips and Learning Wrap-ups.

> *As the teacher you can assess your students' capabilities and perhaps set a quicker goal, or have students set their own goals. Occasionally 40 seconds is too fast. Make sure your goals are appropriate.*

DAY3 1,2,3,10

RAPID WRITING
EXERCISE
WORKSHEET

NAME _____ #

$6 \times 1 =$ _____ $5 \times 2 =$ _____ $11 \times 3 =$ _____ $4 \times 10 =$ _____

$1 \times 1 =$ _____ $6 \times 2 =$ _____ $9 \times 3 =$ _____ $6 \times 10 =$ _____

$8 \times 1 =$ _____ $12 \times 2 =$ _____ $2 \times 3 =$ _____ $5 \times 10 =$ _____

$3 \times 1 =$ _____ $1 \times 2 =$ _____ $3 \times 3 =$ _____ $7 \times 10 =$ _____

$7 \times 1 =$ _____ $2 \times 2 =$ _____ $7 \times 3 =$ _____ $2 \times 10 =$ _____

$10 \times 1 =$ _____ $8 \times 2 =$ _____ $5 \times 3 =$ _____ $11 \times 10 =$ _____

$5 \times 1 =$ _____ $11 \times 2 =$ _____ $12 \times 3 =$ _____ $10 \times 10 =$ _____

$12 \times 1 =$ _____ $3 \times 2 =$ _____ $8 \times 3 =$ _____ $3 \times 10 =$ _____

$2 \times 1 =$ _____ $9 \times 2 =$ _____ $6 \times 3 =$ _____ $8 \times 10 =$ _____

$11 \times 1 =$ _____ $7 \times 2 =$ _____ $10 \times 3 =$ _____ $12 \times 10 =$ _____

$9 \times 1 =$ _____ $4 \times 2 =$ _____ $4 \times 3 =$ _____ $9 \times 10 =$ _____

$4 \times 1 =$ _____ $10 \times 2 =$ _____ $1 \times 3 =$ _____ $1 \times 10 =$ _____

time _____   time _____   time _____   time _____

Wrapping Time   3's ____ sec.   10's ____ sec.

DAY4 3,4,11

RAPID WRITING
EXERCISE
WORKSHEET

NAME _____ #

$11 \times 3 =$ _____ $11 \times 4 =$ _____ $6 \times 11 =$ _____

$9 \times 3 =$ _____ $9 \times 4 =$ _____ $1 \times 11 =$ _____

$2 \times 3 =$ _____ $3 \times 4 =$ _____ $8 \times 11 =$ _____

$3 \times 3 =$ _____ $4 \times 4 =$ _____ $3 \times 11 =$ _____

$7 \times 3 =$ _____ $10 \times 4 =$ _____ $7 \times 11 =$ _____

$5 \times 3 =$ _____ $5 \times 4 =$ _____ $10 \times 11 =$ _____

$12 \times 3 =$ _____ $7 \times 4 =$ _____ $5 \times 11 =$ _____

$8 \times 3 =$ _____ $2 \times 4 =$ _____ $12 \times 11 =$ _____

$6 \times 3 =$ _____ $8 \times 4 =$ _____ $2 \times 11 =$ _____

$10 \times 3 =$ _____ $12 \times 4 =$ _____ $11 \times 11 =$ _____

$4 \times 3 =$ _____ $6 \times 4 =$ _____ $9 \times 11 =$ _____

$1 \times 3 =$ _____ $1 \times 4 =$ _____ $4 \times 11 =$ _____

time _____   time _____   time _____

Wrapping Time   3's ____ sec.   4's ____ sec.

MULTIPLYING BY ELEVEN

11
x1

| 1 | 2 | 3 | | | | | | | | |

Fill in the squares. Write #4 here, and continue. Each time both digits are the same number, you have a multiple of 11. When you get to 10 X 11, use the 10 rule to get your answer.

11
x2

| 12 | | | | | | | | | | |

11
x3

| 23 | | | | | | | | | | |

11
x4

| 34 | | | | | | | | | | |

11
x5

| 45 | | | | | | | | | | |

11
x6

| 56 | | | | | | | | | | |

11
x7

| 67 | | | | | | | | | | |

11
x8

| 78 | | | | | | | | | | |

11
x9

| 89 | | | | | | | | | | |

11
x10

| 100 | | | | | | | | | | |

RAPID WRITER EXERCISE X11 WORKSHEET

1 x 11 = ____   9 x 11 = ____

5 x 11 = ____   3 x 11 = ____

11 x 11 = ____   12 x 11 = ____

4 x 11 = ____   1 x 11 = ____

10 x 11 = ____   6 x 11 = ____

6 x 11 = ____   11 x 11 = ____

9 x 11 = ____   8 x 11 = ____

3 x 11 = ____   2 x 11 = ____

12 x 11 = ____   7 x 11 = ____

8 x 11 = ____   4 x 11 = ____

2 x 11 = ____   5 x 11 = ____

7 x 11 = ____   10 x 11 = ____

time _____

JUST FOR FUN!
AN EXTENDED ACTIVITY

To multiply any two digit number by 11, simply write the two digit number with a space between the two numbers. Add the two digits and write the total in the space in the middle. (column 1)

Example: 24 X 11=

Write: 2   4

Think : 2 + 4 = 6

Between the 2 and 4 or in the tens column,
Write: 6

Thus: 24 X 11 = 264

"TOUGHER STUFF"

If the total for the number in the middle is 2 digits, carry the number in the hundreds column and add. (column 2)

Example:
48 X 11 = ___

Write: 4   8

Think: 4+8=12

Write:
```
   1
 + 428
 ─────
  528
```
Put the 2 between the 4 and 8. Carry the 1.

Add 1+4=5

Therefore: 48 X 11 = 528. Try to *think* the answer without writing the problem.

Column 1	Column 2
24 x 11 = ____	29 x 11 = ____
35 x 11 = ____	37 x 11 = ____
44 x 11 = ____	48 x 11 = ____
61 x 11 = ____	65 x 11 = ____
40 x 11 = ____	48 x 11 = ____
18 x 11 = ____	38 x 11 = ____
54 x 11 = ____	56 x 11 = ____
81 x 11 = ____	84 x 11 = ____
33 x 11 = ____	38 x 11 = ____
27 x 11 = ____	47 x 11 = ____
16 x 11 = ____	66 x 11 = ____
52 x 11 = ____	59 x 11 = ____
90 x 11 = ____	92 x 11 = ____

4X1= _____

1X4= _____

4X2= _____

2X4= _____

4X3= _____

3X4= _____

4X4= _____

4X5= _____

5X4= _____

4X6= _____

6X4= _____

4X7= _____

7X4= _____

4X8= _____

8X4= _____

4X9= _____

9X4= _____

4X10= _____

10X4= _____

4X11= _____

11X4= _____

4X12= _____

12X4= _____

| 4 x1 | 1 x4 | 2 x4 | 4 x2 | 3 x4 | 4 x3 | 4 x4 | 5 x4 | 4 x5 | 6 x4 | 4 x6 |

| 4x7= ___ | 4x8= ___ | 9x4= ___ | 4x10=___ | 4x11=___ | 4x12=___ |
| 7x4= ___ | 8x4=___ | 4x9= ___ | 10x4=___ | 11x4=___ | 12x4=___ |

1. Andra had 4 boxes. Each box had 6 peanuts in it.
   Draw a picture showing how many boxes with
   peanuts Andra had. Write a multiplication
   problem and answer to show how many peanuts
   there were.

The first box is done for you!

_____

2. Jeff collects rocks. He keeps them all in 6
   glass tubes. There are 4 rocks in each tube.
   Draw a picture showing Jeff's rocks. Write
   a multiplication problem and answer to
   show how many rocks he has in all.

The first tube is done for you!

_____

3. Mollie kept her hair barrettes in plastic bags.
   She had 4 bags and each one held 9 barrettes.
   Draw a picture showing her bags and barrettes.
   Write a multiplication problem and answer to
   show how many barrettes in all.

The first bag is done for you!

_____

The first fish in each row is done for you!

4. Henry had some lazer fish stickers.
   There were 9 rows with 4 stickers in each
   row. Draw a picture of Henry's stickers.
   Write a multiplication problem and answer
   to show how many stickers he had in all.

_____

5. Write all of the 4X ___ = ___ problems
   and their commutative
   partners   ___ X 4 =____ .

NUMBER FAMILY FUNSHEET
__ X4 or 4X __
SKIP COUNTING

When we count by multiples of 4, we can also say that we are skip counting. When we skip count, we are saying the numbers or products that belong to a certain number family. Skip count through the number maze. Go around each number as you come to it. Do not cross over a path more than once.

Circle the numbers that belong to the X4's family.

2
(4)
6
8
10
12
14
16
18
20
22
24
26
28
30
32
34
36
38
40
42
44
46
48

Skip count along the SQUARES path. →

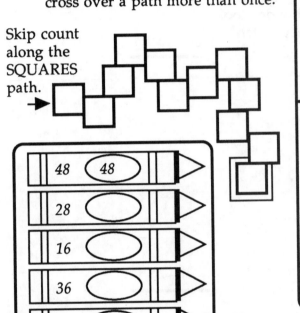

48 (48)
28 ( )
16 ( )
36 ( )
12 ( )
8 ( )
32 ( )
20 (20)
44 ( )
40 ( )
24 ( )
4 (4)

The 4, 20, and 48 crayons are numbered correctly. Can you put the X4 family numbers in order and write them in the ovals?

Start with the 4 and draw lines to the X4, products in order.

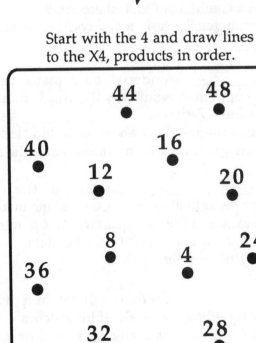

44 48
40   16
12   20
8   4   24
36
32   28

WATCHING T.V. Multiply by 4

Answers   Show what you were
                           thinking to find the answer.

1. _____ watched
TV 4 times this week. He only got to watch
it for 5 minutes each time. How many
minutes did he watch TV this week?   _____

2. _____ and her mother
rented 4 videos last month. Each video took
2 hours. How many hours did they watch
videos last month?   _____

3. _____
kept track of the
commercials on his
favorite program.
There were 8.
It seemed as if
each one lasted 4
minutes. How
many minutes of
commercials would
that have been?   _____

4. If _____ eats 4 cookies during
each commercial and there are 9
commercials, how many cookies would he
eat?   _____

5. _____wondered how many times
the newsman would say the word "today". He
counted 7 times. _____ asked if
the newsman did 4 shows and said "today" 7
times each, how many times would he say it? _____

6. _____ watched the
Jazz basketball game.   Karl Malone made
4 baskets in the first quarter. At 2 points
each, how many points did he make
the first quarter?   _____

7. _____ mother will let him watch
TV for 4 hours a week. If he watches
for 4 weeks, how many hours will he watch?   _____

ARRAY PATTERN

- Arrange the manipulatives so that the product (answers) can be quickly identified.
- Place manipulatives in rows and columns to show the following problems:

  4 X 3, 3 X 4   5 X 2, 2 X 5   8 X 3, 3 X 8
  2 X 6, 6 X 2   7 X 6, 6 X 7   5 X 8, 8 X 5

- Check with your neighbor each time to see if you came up with the same arrangement.

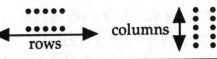

DAY 4

CHECKING PROGRESS

- **Collect parent signed homework slips. Draw for prize(s).**
- **Do the #3 and #4 Wrap-ups. Time and record on the CHECK YOUR PROGRESS CHART.**
- **Do Day 4 RAPID WRITING EXERCISE WORKSHEET (previously printed with Day 3, p. 44)**
- Do not collect the Wrap-ups. They will go home again tonight to practice the #5 board.
- **Have students mark their HELPER CHARTS both down the 3 and 4 columns, as well as across for each of the commutative partners.**

TODAY'S LESSON #5 and Perfect Squares

- **Talk about counting by 5. Name things we use every day that comes in 5's such as nickels, clocks with 5 minute segments, etc.**
- **Do SKIP COUNTING X5 audio with music. Have a student make up and lead exercises.**
- **Do NUMBER FAMILY X5 FUNSHEET.**

- **Practice with #5 Wrap-up.**
- **Do FIELD DAY AT SCHOOL X5 WORKSHEET**

  Other X5 materials available
  > ARRAY PATTERN for use with manipulatives
  > CONCENTRATION X5 DRILL STRIP WORKSHEET
  > SKIP COUNTING X5 EXERCISE video
  > WRAP-UP RAP X5 audio

- **TALK ABOUT perfect squares. On each of the COMMUTATIVE WORKSHEETS the perfect squares have been designated with a bold square because they have no commutative partner. Point out that students have already learned the perfect squares 1X1, 2X2, 3X3, 4X4, 5X5, 10X10 and 11X11. That is more than half of the perfect squares.**

  There are no individual Learning Wrap-ups for perfect squares, but it is important for students to understand them.

- **Do the pages for Perfect Squares, part 1 (p.57) and part 2 (p.58) if necessary.**

- **Do the WORKSHEET (p.59). Students do not need to record their time. Put students in groups of two or three and play TOP THE SQUARES. (p. 60)**

SEND HOME

- Learning Wrap-ups. Ask the parents to time the students as they wrap the #3, #4, and #5 boards without stopping. Parents can check the answer lines. *Tell them there will be a race tomorrow in class.*

- Perfect Square Drill Strips. Ask the parents to call out the drill strips and have the student give the answer.

> TC *Ask the lunchroom cooks to bake cookies for your party next week. (p.101)*

NUMBER FAMILY FUNSHEET
X5 or 5X
SKIP COUNTING

The number family for 5 is also called the product of the factors 5 multiplied by ____. Fill in the factor that has not been written.

5 X		= 5
5 X		= 20
5 X		= 60
5 X		= 15
5 X		= 25
5 X		= 55
5 X		= 40
5 X		= 10
5 X		= 30
5 X		= 35
5 X		= 50
5 X		= 45

PENTAGON TRAIL

Write the numbers as you skip count by 5 down the trail.

Start with the 5 and draw lines to the X5 products in order. Use only the members of the 5's number family.

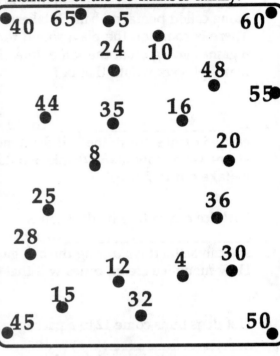

40   65   5   60

24   10

48

44   35   16   55

8   20

25   36

28

12   4   30

15   32

45   50

When we count by multiples of 5, we can also say that we are skip counting. When we skip count, we are saying the numbers or products that belong to a certain number family. Skip count through the number maze. Circle each number as you come to it. Try to do it without lifting your pencil off the paper.

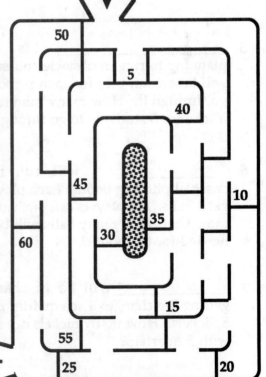

50   5   40   45   35   30   60   10   15   55   25   20

FIELD DAY AT SCHOOL Multiply by 5

Our class is planning to help with Field Day. These are some of the things we have been talking about.

Answers

Show what you were thinking to find the answer.

1. _____ made up a game called pentagon ball. It takes 5 players. There is room on the playground to have 8 games going on at the same time. How many players would that be?

2. If _____ could carry 5 drinks out to the ball diamond at one time, how many drinks would he take out in 7 trips?

3. Each ice cream tray holds 9 cones. _____thinks she can sell 5 trays during the ball game. How many ice cream cones will that be?

4. Hot dogs buns come 12 to a package. _____ says that each grade would need 5 packages. How many buns would that be?

5. _____ is planning to run an obstacle course. It will take 4 minutes for each person to go through it. How many minutes will it take for 5 students to go through?

6. _____ will run the baseball pitching booth. Each player will get 3 balls. If 5 players can pitch at the same time, how many balls will be needed for the booth?

7. _____will be in charge of getting the change. Each quarter is worth 5 nickels. How many nickels can she get with 5 quarters?

5X1= _____	**1X5=** _____
5X2= _____	**2X5=** _____
5X3= _____	**3X5=** _____
5X4= _____	**4X5=** _____
5X5= _____	**5X5=** _____
5X6= _____	**6X5=** _____
5X7= _____	**7X5=** _____
5X8= _____	**8X5=** _____
5X9= _____	**9X5=** _____
5X10= _____	**10X5=** _____
5X11= _____	**11X5=** _____
5X12= _____	**12X5=** _____

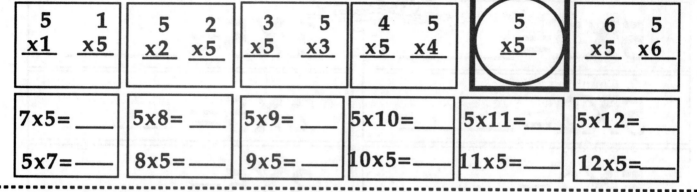

| 5 x1 | 1 x5 | 5 x2 | 2 x5 | 3 x5 | 5 x3 | 4 x5 | 5 x4 | 5 x5 | 6 x5 | 5 x6 |

7x5=___   5x8=___   5x9=___   5x10=___   5x11=___   5x12=___

5x7=___   8x5=___   9x5=___   10x5=___   11x5=___   12x5=___

1. Stephanie counted the windows in the office building where her mother worked. There were 6 windows on each floor and there were 5 floors. Draw a picture of the windows. Write a multiplication problem and answer to show how many windows there were.

One window on each floor has been done for you!

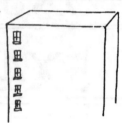

_____

2. Ryan put baseball cards in a book. There were 5 pages with 6 cards on each page. Draw a picture of the pages and cards. Write a multiplication problem and answer to show how many cards Ryan has.

The first page is already done!

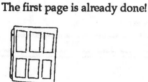

_____

3. Daniel collected sea shells on his vacation. When he lined them up, there were 5 rows with 7 shells in each row. Draw a picture of Daniel's shells. Write a multiplication problem and answer telling how many shells Daniel has.

One row of shells is already done!

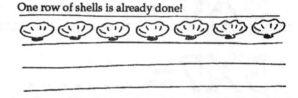

_____

4. When Kemper's mom took cookies out of the oven, Kemper arranged them so there were 7 columns with 5 cookies in each column. Draw a picture of the cookies. Write a multiplication problem and answer to show how many cookies there were.

The top cookie in each column is already done!

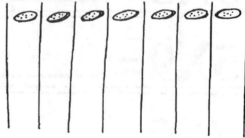

5. Write all of the 5X ___ = ___ problems and their commutative partners ___ X 5 =____.

PERFECT SQUARES

1. Cut out each of the dark outlined boxes on the next page.

2. Count the number of small squares in each box and write the number in the lower right hand corner. You now have the answers to the perfect square multiplication problems for each box.

3. Starting with 1X1, then 2X2, etc. lay each box on the grid below, then lift the corner with the squared number and write the answer in the dark outlined square under it on the grid.

4. Finish by using what you know about the 10 and the 11 rules. The answer to 12 X 12 is hiding somewhere on this page. Circle it and write it in the lower right hand corner.

A PERFECT SQUARE
is as wide as it is tall.

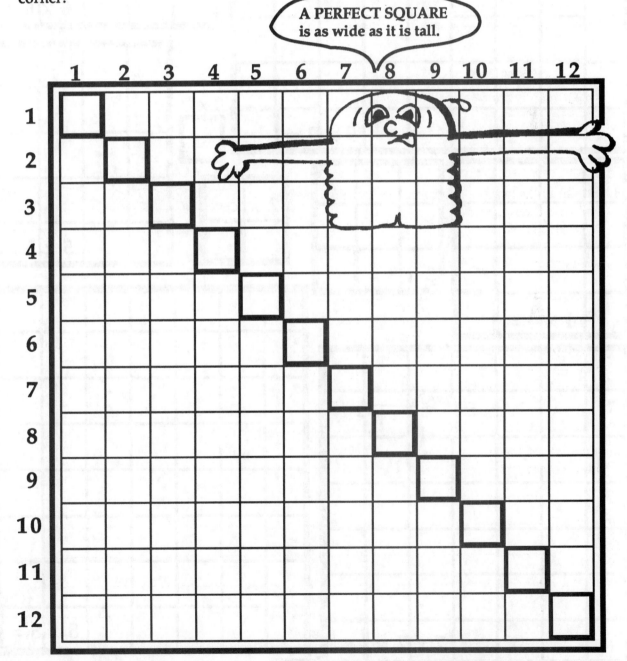

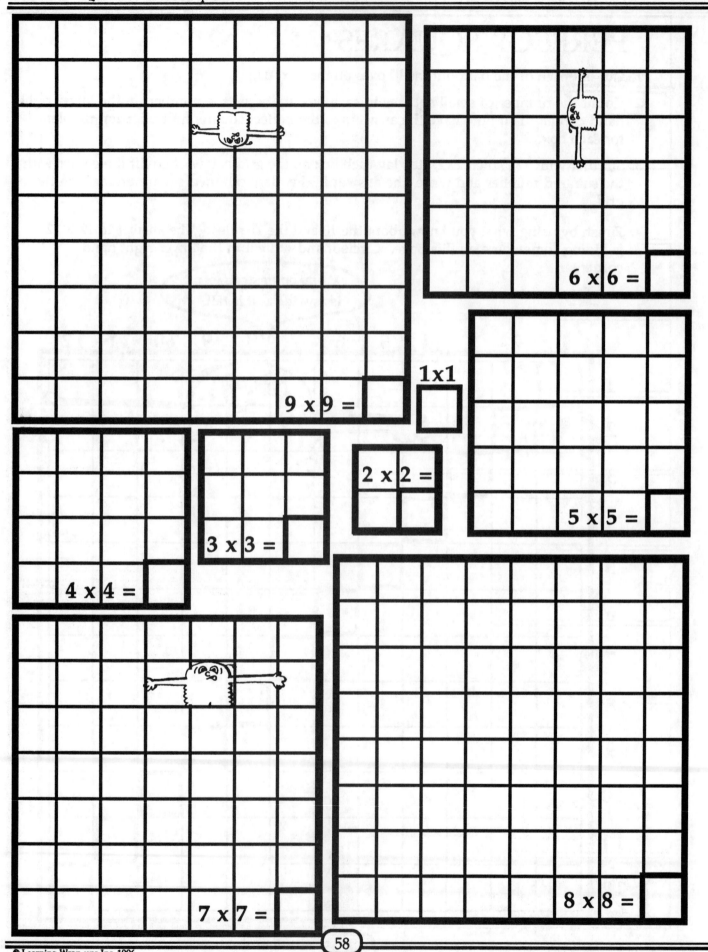

9 x 9 =

1x1

6 x 6 =

2 x 2 =

5 x 5 =

3 x 3 =

4 x 4 =

7 x 7 =

8 x 8 =

1X1=_____

2X2=_____

3X3=_____

4X4=_____

5X5=_____

6X6=_____

7X7=_____

8X8=_____

9X9=_____

10X10=_____

11X11=_____

12X12=_____

Write the answers as fast as you can.

1 X 1 = _____	3 X 3 = _____
2 X 2 = _____	7 X 7 = _____
3 X 3 = _____	11 X 11 = _____
4 X 4 = _____	4 X 4 = _____
5 X 5 = _____	6 X 6 = _____
6 X 6 = _____	1 X 1 = _____
7 X 7 = _____	8 X 8 = _____
8 X 8 = _____	12 X 12 = _____
9 X 9 = _____	5 X 5 = _____
10 X 10 = _____	2 X 2 = _____
11 X 11 = _____	9 X 9 = _____
12 X 12 = _____	10 X 10 = _____

Write the factors for these perfect squares.

64= _____	121 = _____
4 = _____	9 = _____
100= _____	144= _____
36 = _____	2 = _____
49 = _____	16 = _____
81 = _____	25 = _____

TOP THE SQUARES

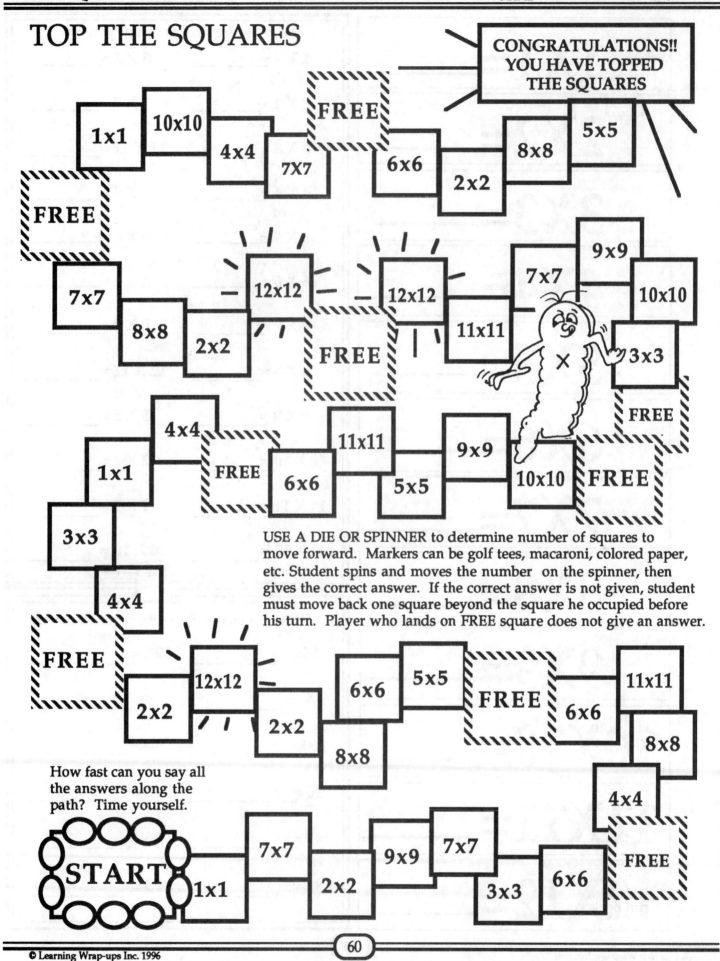

CONGRATULATIONS!!
YOU HAVE TOPPED
THE SQUARES

1x1   10x10   4x4   7X7   FREE   6x6   2x2   8x8   5x5

FREE

7x7   8x8   2x2   12x12   12x12   11x11   7x7   9x9   10x10   3x3

FREE   FREE

4x4   1x1   FREE   6x6   11x11   5x5   9x9   10x10   FREE

3x3

4x4

USE A DIE OR SPINNER to determine number of squares to
move forward. Markers can be golf tees, macaroni, colored paper,
etc. Student spins and moves the number on the spinner, then
gives the correct answer. If the correct answer is not given, student
must move back one square beyond the square he occupied before
his turn. Player who lands on FREE square does not give an answer.

FREE   2x2   12x12   2x2   6x6   5x5   FREE   6x6   11x11   8x8

8x8   4x4

How fast can you say all
the answers along the
path? Time yourself.

START   1x1   7x7   2x2   9x9   7x7   3x3   6x6   FREE

DAY 5

CHECKING PROGRESS

- Collect parent-signed homework slips. Draw for prize. Praise!

- Have students practice wrapping the #3, #4, and #5 Learning Wrapups as fast as they can without stopping between boards. Time them. This is called a Three Board WRAP-OFF.

> *NOTE: Students usually need to stand to do a Wrap-off. When they finish, have them sit down.   Call "TIME" in two minutes, then have all students sit down. (This keeps students from being embarrassed if they are slow, but gives incentive to work harder.)*

- Do the Day 5 RAPID WRITING EXERCISE WORKSHEET for 3, 4, 5 and Perfect Squares. Allow 40 seconds for each column. Time the students so they can see and hear you count. If they do not finish in the 40 seconds, have them underline the last problem they completed in the time limit, then complete the column.

- Fill in HELPER CHART. Have students count the facts they have already learned. Praise!

- Set aside the Learning Wrap-ups. They will be used to practice the #6 Wrap-up, today.

TODAY'S LESSON #6

- Pass out HERE'S WHAT I'VE LEARNED CHART (p.16 section 2) to each student. Have students write the answers to **only the perfect square problems** which they learned yesterday. Color each of the squares they have done correctly with a purple pencil or crayon.

- Ask students to write the answers to all of the X6 and 6X problems that they have learned so far. Encourage them to remember they have learned 1X6, 6X1, 2X6, 6X2, 3X6, 6X3, 4X6, 6X4, 5X6, 6X5, 6X6, 10X6, 6X10, 11X6, 6X11. Have them count the number of X6 problems they know and color them with a blue pencil or crayon.

- Write these four problems and answers on the board. 7X6 = 42, 8X6 = 48, 9X6 = 54, 12X6 = 72. Leave this information on the board temporarily. Have students write the answers on their HERE'S WHAT I'VE LEARNED CHART, but do not color it yet.

- TALK ABOUT when a number is multiplied by 3 then doubled, it has the same answer as that particular number multiplied by 6. Examples: 3 X 7 = 21, doubled, it equals 42. 6 X 7 = 42. Another example, 3 X 8 = 24, doubled, it equals 48. 6 X 8 = 48. 3 X 9 = 27, 27 doubled is 54. 6 X 9 = 54.

- Play the WRAP-UP RAP X6 AUDIO two times. Let Students move about while wrapping the Wrap-up and calling out the answers.

- Do the NUMBER FAMILY X6 FUNSHEET.

- Put students in teams of three to five people. Have them race to front of class room, write the answers on one side of their own CONCENTRATION X6 DRILL STRIP WORKSHEET, run back and tag the next person.

Day 5 (cont)

- Work with the #6 Learning Wrap-up. Have students determine which problems they have not yet learned (7x6, 8x6, 9x6, 12x6). Count them - only four.

- **Drill with the #6 Learning Wrap-ups.**

- **Use the counting cards for 6 CHICKS instead of manipulatives. Multiples of 6 quantities can be miscounted or spilled. Using the counting cards is more convenient and students can count the chicks instead of individual manipulatives if necessary.**

  To play with 6 CHICKS, one student puts down a number of cards, counting them as each card is put down (#1, #2, #3, etc). Another student tells quickly how many chicks total on the number of cards put down. Encourage students to think mentally about the total number of chicks each time a card is put down. This is skip counting. Also encourage them to use the larger numbers for their practice. They could start by putting down five cards, then quickly add or take away more cards.

- **Have students think about 6 packs of soda pop. Use the blank cards (p. 104 section 4) in the extended activities section for students to draw cans of soda. Talk about how many cans in 1 case and 2 cases. If you have community support, this is a good day to bring in the real thing.**

> *Research indicates that we remember what is funny, unusual, odd or bizarre. Students need all the help they can get to remember the 6 times tables.*

- Do an art project. Have students pick an animal from the FUNNY FARM INDIVIDUALIZED STORY PROBLEMS WORKSHEET and do a large painting or drawing. Or have them fold an 11" by 17" paper into 8 sections and create an animal for each of the problems 2 through 9.

  Other X6 materials available
  COMMUTATIVE X6 WORKSHEET
  SKIP COUNTING EXERCISE X6 VIDEO
  WRAP-UP RAP X6 AUDIO

- **It is still important for the teacher to keep moving about the class to spot any problems a student might be having.**

SEND HOME

- Learning Wrap-ups. Practice the #3, #4, and #6 boards. Tell students to do the #6 board 21 times. They can mark on the back of their homework practice slip every time they do it so they will be sure to do it enough times.

  Have students ask their parents to time the first time and the last time. Tell them you will be having a 3, 4, and 6 WRAP-OFF on Day 6. Challenge them to wrap all 3 boards in 1 1/2 minutes, or your appropriate goal time.

DAY 5 3, 4, 5 PERFECT SQUARES

NAME _____ # _____

RAPID WRITING EXERCISE WORKSHEET

11 X 3 = ___	11 X 4 = ___	6 X 5 = ___	12 X 12 = ___
9 X 3 = ___	9 X 4 = ___	1 X 5 = ___	7 X 7 = ___
2 X 3 = ___	3 X 4 = ___	8 X 5 = ___	8 X 8 = ___
3 X 3 = ___	4 X 4 = ___	3 X 5 = ___	6 X 6 = ___
7 X 3 = ___	6 X 4 = ___	7 X 5 = ___	10 X 10 = ___
5 X 3 = ___	5 X 4 = ___	10 X 5 = ___	4 X 4 = ___
12 X 3 = ___	7 X 4 = ___	5 X 5 = ___	1 X 1 = ___
8 X 3 = ___	2 X 4 = ___	12 X 5 = ___	5 X 5 = ___
6 X 3 = ___	8 X 4 = ___	2 X 5 = ___	11 X 11 = ___
10 X 3 = ___	12 X 4 = ___	11 X 5 = ___	9 X 9 = ___
4 X 3 = ___	10 X 4 = ___	9 X 5 = ___	2 X 2 = ___
1 X 3 = ___	1 X 4 = ___	4 X 5 = ___	3 X 3 = ___

time ___ time ___ time ___ time ___

Wrapping Time #3 ___ sec. #4 ___ sec. #5 ___ sec.

DAY 6 4, 6, 10, 7

NAME _____ # _____

RAPID WRITING EXERCISE WORKSHEET

Think about commutative partner 7's and write all the answers you know. →

11 X 4 = ___	8 X 6 = ___	6 X 10 = ___	11 X 7 = ___
9 X 4 = ___	4 X 6 = ___	3 X 10 = ___	9 X 7 = ___
3 X 4 = ___	10 X 6 = ___	8 X 10 = ___	2 X 7 = ___
4 X 4 = ___	2 X 6 = ___	10 X 10 = ___	3 X 7 = ___
10 X 4 = ___	1 X 6 = ___	2 X 10 = ___	7 X 7 = ___
5 X 4 = ___	7 X 6 = ___	5 X 10 = ___	5 X 7 = ___
7 X 4 = ___	5 X 6 = ___	1 X 10 = ___	12 X 7 = ___
2 X 4 = ___	9 X 6 = ___	9 X 10 = ___	8 X 7 = ___
8 X 4 = ___	11 X 6 = ___	4 X 10 = ___	6 X 7 = ___
12 X 4 = ___	3 X 6 = ___	12 X 10 = ___	10 X 7 = ___
6 X 4 = ___	12 X 6 = ___	7 X 10 = ___	4 X 7 = ___
1 X 4 = ___	6 X 6 = ___	11 X 10 = ___	1 X 7 = ___

time ___ time ___ time ___ time ___

Wrapping Time #4 ___ sec. #6 ___ sec. #10 ___ sec.

NUMBER FAMILY FUNSHEET
___ X6 or 6X ___
SKIP COUNTING

Write the numbers that belong to the X6 family in the HEXAGONS.

6
12
18
24
30
36
42
48
54
60
66
72

Skip count through the X6 number family four times. Each time you get to 72 start over with 6. Draw a path with your pencil.

Start here.

6

This Silly Six Snake is about to shed its skin. It can keep only that skin which is numbered a multiple of 6. Help it by coloring the spots that are not multiples of 6.

6666 6 6   6   4   9   10   12   14   15   17   18   21   22   24   25   27   28   30   32   36   37   40   42   44   45   48   53   54   55   56   57   60   63   65   66   70   72   77   79   80

6   12   24   18   30   66   54   36   60   72   48   42   66   6   18   60   12   24   72   54   48   42   30   6   54   48   36   12   18   24   30   36   48   42   36   72 End   54   60   66   6   12   66   60   18   72   42   54   24   30   36   48

Color the bricks that belong to the X6 family YELLOW and all others red.

	6	15	24	82	66	75	45	32	17	29	36	40	48	21
19	36	72	60	19	24	33	18	6	54	36	15	38	30	
	6	72	12	60	18	54	24	48	30	42	36	12	6	18
22	24	36	47	56	60	18	66	33	77	72	54	36	30	
	54	18	12	60	72	66	54	30	36	24	18	6	12	42
56	16	14	58	21	63	58	28	44	24	38	40	23	76	

$6X1=$ _____

$1X6=$ _____

$6X2=$ _____

$2X6=$ _____

$6X3=$ _____

$3X6=$ _____

$6X4=$ _____

$4X6=$ _____

$6X5=$ _____

$5X6=$ _____

$6X6=$ _____

$6X6=$ _____

$6X7=$ _____

$7X6=$ _____

$6X8=$ _____

$8X6=$ _____

$6X9=$ _____

$9X6=$ _____

$6X10=$ _____

$10X6=$ _____

$6X11=$ _____

$11X6=$ _____

$6X12=$ _____

$12X6=$ _____

LET'S BUILD A FUNNY FARM Multiply by 6

If our class could work on the funny farm when we finished our assignment, we would build all kinds of strange, funny looking animals for it. Here are some of the things you might see. Draw pictures of some of the animals for an art project.

Answers

Show how you were thinking to find the answers.

1. _____ made an animal with 6 arms. Each arm had 2 hands. How many hands in all?

_____

2. _____ and _____ worked on an animal they called a "tailaroo". It had 6 tails. Each tail had 5 rings on it. How many rings were there in all?

_____

3. _____ helped _____ make teeth for her 6 mouthed animal. Each mouth had 8 teeth. How many teeth did the animal have?

_____

4. Since the animals needed a place to stay, _____ and _____ built a barn. Their barn had 4 sides and each side had 6 windows. How many windows did the barn have?

_____

5. The animal _____ and _____ made had 6 heads. Each head had 3 eyes. How many eyes were there in all?

_____

6. _____ and _____ made an animal with a sign on its chest. There were 4 lines on the sign. Each line had 6 words. How many words were there ?

_____

7. _____'s and _____'s animal was covered with spots. There were 6 big spots. But each big spot had 9 little spots. How many little spots were there all together?

_____

8. _____ and _____ called their animal "eyebrow-brow". It had 7 eyebrows over each eye. It had 6 eyes. How many eyebrows?

_____

9. _____ said there were 10 animals with 6 legs each. How many legs?

_____

1 6	2 6	3 6	4 6	5 6	6
x6 x1	x6 x2	x6 x3	x6 x4	x6 x5	x6

6x7=___	6x8=___	6x9=___	6x10=___	11x6=___	6x12=___
7x6=___	8x6=___	9x6=___	10x6=___	6x11=___	12x6=___

1. Keni helped in the library. She put 6 books on each shelf. There were 4 shelves. Draw a picture of the books and shelves. Write a multiplication problem and answer to show how many books Keni put away.

The first two shelves are done!

_____

2. Taylor put a dime collection in rows and columns. There were 6 columns with 4 dimes in each column. Draw a picture to show how the dimes were arranged. Write a multiplication problem and answer to show how many dimes Taylor had in all.

The first dime is in each column.

_____

3. Amy wanted to freeze some snowballs until next summer. The box she found held 5 rows, with 6 snowballs in each row. Draw a picture of her box of snowballs. Write a multiplication problem and answer to tell us how many snowballs she could save in the box.

The first row of snowballs is done.

_____

4. Bob and Carolyn stacked the math books for the teacher. They had 5 stacks with 6 books in each stack. Draw a picture of the math books when they were finished. Write a multiplication problem and answer to show how many books they stacked.

The last stack is done!

MATH
MATH
MATH
MATH
MATH

MATH MATH MATH MATH MATH

_____

5. Write all of the 6X ___ = ___ problems and their commutative partners ___ X 6 = ___.

_____ _____ _____
_____ _____ _____
_____ _____ _____
_____ _____ _____
_____ _____ _____

DAY 6

CHECKING PROGRESS

- Collect homework slips. Draw for prize(s). Praise!

- Do the X6 Learning Wrap-up. Time and record.

- Practice the #3, #4, and #6 Learning Wrap-ups. Have a WRAP-OFF. Allow two or three minutes (whichever is appropriate for your class) or until half of the class is sitting down, then call time.

- Do Day 6 RAPID WRITING EXERCISE (copied with Day 5). Allow 30 seconds for each column except the #7. Students will notice that they have not done the #7 Learning Wrap-up or writing exercises. Acknowledge the fact, but remind them that they know all but three of the commutative partners. Tell them to turn the problems around if it will help and write all of the answers they can think of. Allow up to one minute for the 7's column.

- Fill in HELPER CHART. Count the fact combinations left to learn. PRAISE! TREAT!

- Talk about what the students have accomplished in the past 5 days. Tell students they have done so well that you are going to teach them one of the easiest sets of facts.

TODAY'S LESSON #9, #7

- Write the X9 facts without the answers on the board. Pass out TEACHING 9'S WORKSHEET and discuss the concept until students understand it.

- Do TEACHING 9'S WORKSHEET.

- Show students how to do the FINGER 9's.   Encourage them to think about it mentally so they do not need to actually use their fingers.

- Do the CONCENTRATION X9 DRILL STRIP WORKSHEET.

- Work with the #9 Learning Wrap-up. Practice for about 10 minutes.

- PRAISE! PRAISE!! This has been a big day. Reward!!!

Other material available for the 9's
    COMMUTATIVE X9 WORKSHEET
    NUMBER FAMILY X9 FUNSHEET
    INDIVIDUALIZED STORY PROBLEMS X9, PLANNING A PARTY WORKSHEET

Day 6 (cont)

• The 7 TIMES TABLES are usually one of the hardest for students to learn. It would be wise to do all of the worksheets provided.

• Turn it into an event! Set up four stations, using your volunteers. As quickly as students finish a page, they give it to the stationmaster for a reward. Mathnique Money, certificates, or treats can be the rewards. Stationmaster gives student the next worksheet, student goes to seat, completes work and delivers it to the next stationmaster.

• Ask stationmasters to correct as many sheets as possible.

• Work sheets to be used:
> CONCENTRATION X7 DRILL STRIP WORKSHEET
> COMMUTATIVE X7 WORKSHEET
> INDIVIDUALIZED STORY PROBLEMS X7 WORKSHEET
>> Fill in the names of friends that are not in your class. Tell students to just work the problems. They do not need to describe **how** they came up with the answer today.
>
> 7 PUPS COUNTING CARDS if needed.

> *The easiest way to remember 7 x 8 is to think 56 = 7 x 8, (#5, #6, #7, #8).*

• As the students finish the paper chase, have them work with the #7 Learning Wrap-up and challenge them to do it in 30 seconds or less.

• Have students give themselves a big cheer!

SEND HOME

• Learning Wrap-ups. Focus on the #7 and #9 boards.

• The worksheets completed and corrected today so the student knows where they stand with the 7 and 9 multiplication problems.

• Remind them to have Wrap-up races with their parents.

• Be sure to bring back the Wrap-ups and homework slips!!!!!

TEACHING X9'S

The 9's are easy. Just learn a couple of strategies.

Think one less than the number you are multiplying by.
Next think of a number that can be added to it which will total 9.
Those two numbers are the answer.

Example:
9 x 4. Think: "1 less than 4 is 3. 3 + 6 = 9. So 36 is the answer."
Try 9 x 8. Think: " 1 less than 8 is 7. 7 + ? = 9.   7 + 2 = 9, so 72 is the answer."

Think....
1 less than, is, plus a number, that will total 9.

9x1=___   0+9=9

9x2=___   1+8=9

9x3=___   2+7=9

9x4=___   3+6=9

9x5=___   4+5=9

9x6=___   5+4=9

9x7=___   6+3=9

9x8=___   7+2=9

9x9=___   8+1=9
                use
9x10=___ the 10 rule
                use
9x11=___ the 11 rule
                memorize
9x12=___ 108

Now zip through the 9's.

9x1=___	10x9=___	12x9=___
9x2=___	3x9=___	2x9=___
9x3=___	9x9=___	4x9=___
9x4=___	6x9=___	8x9=___
9x5=___	1x9=___	6x9=___
9x6=___	7x9=___	1x9=___
9x7=___	11x9=___	10x9=___
9x8=___	5x9=___	3x9=___
9x9=___	2x9=___	9x9=___
9x10=___	4x9=___	7x9=___
9x11=___	8x9=___	11x9=___
9x12=_108_	12x9=___	5x9=___

If students don't pick up on the "1 less" system for multiplying by 9, you may want to use this old trick. Kids love it, but sometimes the fingers take longer than desirable to solve the problem. If you use this, be sure to have students practice visualizing the fingers rather than actually putting them out on their desks.

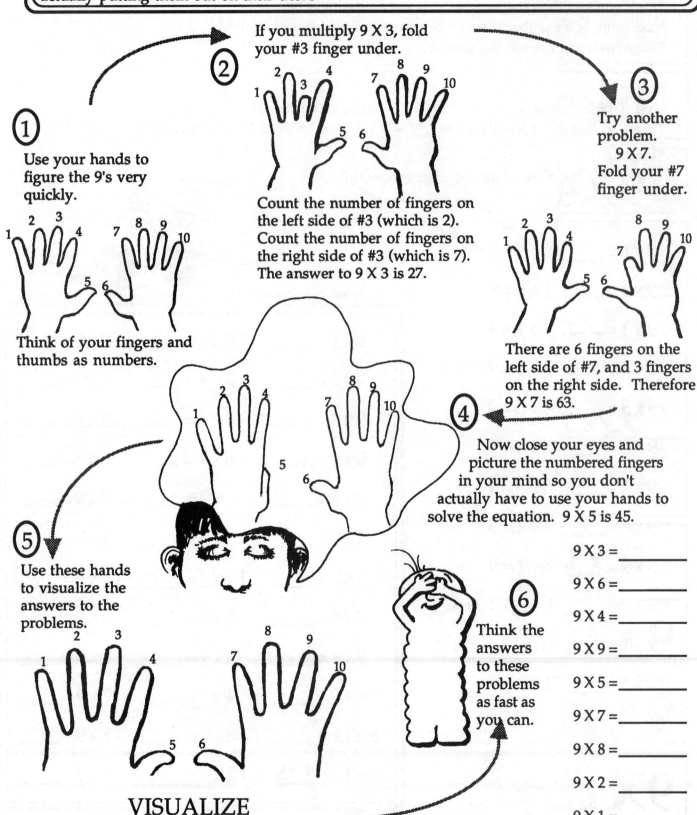

① Use your hands to figure the 9's very quickly.

Think of your fingers and thumbs as numbers.

② If you multiply 9 X 3, fold your #3 finger under.

Count the number of fingers on the left side of #3 (which is 2). Count the number of fingers on the right side of #3 (which is 7). The answer to 9 X 3 is 27.

③ Try another problem. 9 X 7. Fold your #7 finger under.

There are 6 fingers on the left side of #7, and 3 fingers on the right side. Therefore 9 X 7 is 63.

④ Now close your eyes and picture the numbered fingers in your mind so you don't actually have to use your hands to solve the equation. 9 X 5 is 45.

⑤ Use these hands to visualize the answers to the problems.

VISUALIZE

⑥ Think the answers to these problems as fast as you can.

9 X 3 = _____
9 X 6 = _____
9 X 4 = _____
9 X 9 = _____
9 X 5 = _____
9 X 7 = _____
9 X 8 = _____
9 X 2 = _____
9 X 1 = _____

72

9X1= _____	1X9= _____
9X2= _____	2X9= _____
9X3= _____	3X9= _____
9X4= _____	4X9= _____
9X5= _____	5X9= _____
9X6= _____	6X9= _____
9X7= _____	7X9= _____
9X8= _____	8X9= _____
9X9= _____	9X9= _____
9X10= _____	10X9= _____
9X11= _____	11X9= _____
9X12= _____	12X9= _____

| 1   9 | 2   9 | 3   9 | 4   9 | 5   9 | 9x6=___ |
| x9   x1 | x9   x2 | x9   x3 | x9   x4 | x9   x5 | 6x9=___ |

| 9x7=___ | 9x8=___ | 9 x9 | 9x10=___ | 9x11=___ | 9x12=___ |
| 7x9=___ | 8x9=___ | | 10x9=___ | 11x9=___ | 12x9=___ |

1. Jackie counted 9 toothpicks and put them in a bundle She made 8 bundles. Draw a picture showing the toothpicks in bundles. Write a multiplication problem and answer to show how many toothpicks Jackie needed for her assignment.

The first bundle is done.

_____

2. Brady was working on the same problem, but he considered the commutative property. He used 8 small plastic cups and put 9 chocolate chips in each one. Draw a picture showing Brady's problem. Write a multiplication problem and answer to show how many chocolate chips he needed in all.

The plastic cups are already drawn.

_____

3. Erin had a box of chalk. There were 3 sticks in a column and there were 9 sticks in a row. Draw a picture showing the chalk in the box. Write a multiplication problem and answer to show how many sticks of chalk the box held.

The bottom row of chalk is already in the box.

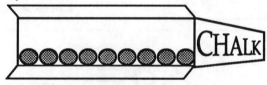

_____

4. Ellen drew enough hearts to make 9 columns with 3 hearts in each column. Draw a picture like Ellen's. Write a multiplication problem and answer showing how many hearts she drew.

The first heart in each column is drawn.

_____

5. Write all of the 9X ___ = ___ problems and their commutative partners ___ X 9 =___.

_____ _____ _____

_____ _____

_____ _____

_____ _____

_____

NUMBER FAMILY FUNSHEET
___ X9 or 9X ___
SKIP COUNTING

Starting with cloud 9, write the numbers of the X9 family in order.

9

108

Write the X9 family numbers two times.

9
18
27
36
45
54
63
72
81
90
99
108

Can you find your way to the center of this maze? You must pass through the X9 family numbers in order. Do not travel any path more than once. Try to do it without lifting your pencil off the paper.

9

63
45
90
27
72
108
99
36
81
18
54

Nonagon 1:
108 29 99 90 81 49 63 69 72 27 36 18 9 54 93 45

Nonagon 2:
108 72 49 18 27 63 45 56 81 54 36 97 90 9 32 99

Nonagon 3:
9 65 81 63 90 54 36 99 42 108 45 18 72 91 27 39

Draw a circle around the four numbers in each NONAGON that are not members of the X9 family.

PLANNING A PARTY Multiply by 9

Some of the students in our class helped plan a party for the kindergarten.
Each of our class members helped with a group of 9 children. Here are
their problems.

Answer

Show how you were
thinking to find the answer.

1. The 9 children in _____ 's
   group had 4 sacks each to make puppets.
   How many sacks did they have?   _____

2. _____ let
   each of the 9 children in her group
   list their 3 favorite games. If all the
   games were different, how many
   games were on the lists?   _____

3. _____
   gave each of the 9 children 10 little
   marshmallows and several
   toothpicks to make monsters. How
   many marshmallows did she give?   _____

4. _____'s
   group of 9 each received a box with
   8 crayons to draw pictures of themselves. What was the total number
   of crayons in the boxes?   _____

5. _____
   planned 2 hot dogs for each of her
   9 kindergarten children. How
   many hot dogs did she need?   _____

6. _____ helped her
   9 students string colored beads for
   bracelets. Each string had 7 beads.
   How many beads did she use?   _____

7. A mother brought cupcakes for
   _____'s group of 9
   children. Each cupcake had 6 candies
   on top. How many candies were there?   _____

8. _____ gave each child
   in her group of 9, 9 jelly beans. How
   many jelly beans did she give?   _____

7X1= _____	1X7= _____
7X2= _____	2X7= _____
7X3= _____	3X7= _____
7X4= _____	4X7= _____
7X5= _____	5X7= _____
7X6= _____	6X7= _____
7X7= _____	7X7= _____
7X8= _____	8X7= _____
7X9= _____	9X7= _____
7X10= _____	10X7= _____
7X11= _____	11X7= _____
7X12= _____	12X7= _____

| 1
x7 | 7
x1 | 2
x7 | 7
x2 | 3
x7 | 7
x3 | 4
x7 | 7
x4 | 5
x7 | 7
x5 | 7x6= ___
6x7= ___ |

| 7
x7 | 7x8= ___
8x7= ___ | 7x9= ___
9x7= ___ | 7x10= ___
10x7= ___ | 7x11= ___
11x7= ___ | 7x12= ___
12x7= ___ |

1. Cassie noticed the February calendar had 4 weeks. There were 7 days in each week. The first day was on Sunday. Write the numbers on the calendar then write a multiplication problem and answer to show how many days there are on the February calendar. It is not a leap year.

The calendar is done. You write in the numbers.

FEBRUARY

| 1 | 2 | | | | | |

This is the first row..

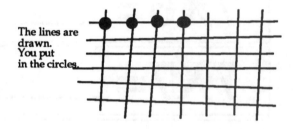

2. Jared hooked paper clips together for a math lesson. He had 7 paper clips in a line and there were 4 lines. Draw a picture showing his paper clips. Write a multiplication problem and answer to show how many paper clips he had in all.

3. Lynnette used lines to help her learn the 7 times tables. She drew 7 columns (lines up and down) and 6 rows (lines going across). Then she drew a circle at every spot where two lines crossed. Show us what Lynnette's picture looked like. Write a multiplication problem and answer to show how many circles she drew.

The lines are drawn. You put in the circles.

You draw the lines and circles.

4. Tex liked the way Lynnette solved her problem, so he decided to draw 6 lines down and 7 lines across. Show us what his drawing looked like. Write a multiplication problem and answer to show how many circles Tex drew where the lines crossed.

5. Write all of the 7X __ = __ problems and their commutative partners ___ X 7 =__.

_____ _____       _____ _____

_____ _____   _____ _____

_____ _____   _____ _____

_____ _____   _____ _____

NUMBER FAMILY FUNSHEET
____ X7 or 7X ____
SKIP COUNTING

Write the X7 number family in order two times. Try to memorize them as you write.

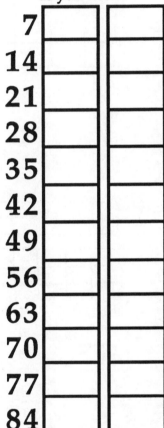

7
14
21
28
35
42
49
56
63
70
77
84

Circle the three numbers in each HEPTAGON that do <u>not</u> belong to the X7 family.

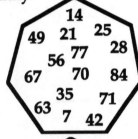

14
49 21 25
56 77 28
67 70 84
35 71
63 7 42

35
84 56 77
17 28 63
21
49 7 72 37
70 42
14

84
21 71 35
56 70
77 28 49
27
47 63 14
7 42

Write the X7 family numbers in order on the line above the dots.

Number Family X7

Factors	Product	Factors
7X1	7	3X7
7X4	14	1X7
7X5	21	4X7
7X2	28	2X7
7X3	35	5X7
7X6	42	7X7
7X8	49	6X7
7X10	56	8X7
7X7	63	12X7
7X12	70	9X7
7X9	77	11X7
7X11	84	10X7

How fast can you draw lines from the factors to the correct product? Two have already been done for you.

1   2   3   4   5   6   7   8   9   10   11   12

READING ABOUT ANIMALS
Multiply by 7

Answers   Show how you were thinking to find the answer.

1. _____ read about 7 kinds of animals that went aboard Noah's Ark. She knew that there were 2 of each kind. How many animals did she read about ? _____

2. In an animal picture book _____ counted 5 animals on each page. There were 7 pages. How many animals were there all together? _____

3. In a book about bears that _____ read, the author told about 3 different types of bears on each page. There were 7 pages. How many types of bears were there? _____

4. _____ chose a joke book about dogs with 6 legs. There were 7 different dogs. How many legs did they have in all? _____

5. _____ could read 4 pages in one minute. How many pages could she read in 7 minutes? _____

6. _____ discovered his book had 7 different animals in each picture. There were 9 pictures. How many animals were there in all? _____

7. The book _____ read was about drawing animals. He could draw a horse in 7 minutes. If he drew 8 horses, how many minutes would it take? _____

8. When _____ reads to the class after lunch, it takes 7 minutes to read a chapter. If she reads 7 chapters, how many minutes will it take? _____

DAY 7

CHECKING PROGRESS

- Collect homework slips. Draw for a prize. Praise!

- Do the #7 and #9 Learning Wrap-ups. Time and record on the TRACKING PROGRESS CHART.

- Do Day 7 RAPID WRITING EXERCISE WORKSHEET. Encourage students to think of commutative partners for the 8's and write as many answers as they can. Have students correct their sheets and turn them in. TELL THEM HOW IMPORTANT IT IS THAT THEY MARK ERRORS. This is the only way you will know if they still need some help. REMIND THEM AGAIN that their life will be much easier if they learn all the facts now, and you cannot help if you do not know they need it.

- Fill in HELPER CHART. Ask how many sets of facts there are left to learn. There is ONLY ONE and its COMMUTATIVE PARTNER!!!!! (8X12, 12X8)

- While students are filling in their charts, check to see how well they did on the RAPID WRITER EXERCISE WORKSHEET.

TODAY'S LESSON #8

- Ask if anyone knows the answer to 8X12. Have them tell you all the different ways they could use to find the answer. Any answer is OK. Suggest the one below if no one mentioned it.

> *An easy way to remember 8X12 is to think 10X8 = 80 + 2X8. (80 + 16 = 96)*

- Tell the students they have learned all of the X8 facts because they have learned all of the X8 commutative partners. Verbally go through the 8 times tables while students call out the answers. Do SKIP COUNTING X8 with the audio or video. Do the WRAP-UP RAP with the Wrap-ups.

- Work with the #8 Learning Wrap-up for the rest of the period.

- Write three goals for the #8 Wrap-up on the chalkboard, 40 seconds, 30 seconds and less than 30. (Younger children may take longer.)

> *Tell the students that you will periodically call up all the people who have reached a particular goal to do it in front of the class. Say, "So be absolutely sure you can do it in the number of seconds you say you can."*

Teacher counts the seconds to 40. Students who have completed their Wrap-ups come to the board and write their names under the appropriate goal. Those who have not finished continue to work until they complete the Wrap-up.

Teacher counts the seconds again. Students who did it in 40 seconds now work for 30. Each time they reach a goal, they write their name on the board.

The teacher continues to count in 40-second segments until class period is over.

DAY 7 (cont)

(The same thing could be done with the ALL-PURPOSE ROLL SHEET found on p. 122 section 5. Write the goals under the 8 at the top of the sheet. Have students report to you instead of writing their names on the board.)

- If students have been keeping Wrap-ups in their desk, have them put the set in order and turn them in. Teacher collects! This way you can tell if any are missing and make the student responsible to get the set together.

SEND HOME

- All of the X8 worksheets. Tell students to work the problems as fast as they can and to have their parents correct them.

  Material available:
       CONCENTRATION DRILL STRIPS X8 WORKSHEET
       COMMUTATIVE X8 WORKSHEET
       NUMBER FAMILY X8 FUNSHEET
       VISITING THE FARM X8 WORKSHEET
       8 CAR CARDS if necessary
       WEB SIGHT MULTIPLICATION #1 and #2 (inside covers)

TEACHER INFORMATION

Make certain all of the Learning Wrap-ups are put into complete sets for students to use on DAY 8. This is a good job for your volunteers if the students have not already done it. They are now ready for the students to have a Wrap-off with the entire set of boards.

Review student materials. Determine which students require extra teaching and determine if and how they should be grouped according to their needs. Assign volunteers a specific time to help the groups.

Day 8 is an important review day. Make certain everything is ready to go.

RAPID WRITING EXERCISE WORKSHEET

Think about the commutative partners of 9 and write all the answers you know. →

4 X 4 = ___	10 X 6 = ___	6 X 7 = ___	6 X 9 = ___
9 X 4 = ___	5 X 6 = ___	10 X 7 = ___	3 X 9 = ___
5 X 4 = ___	12 X 6 = ___	4 X 7 = ___	8 X 9 = ___
7 X 4 = ___	1 X 6 = ___	1 X 7 = ___	10 X 9 = ___
2 X 4 = ___	2 X 6 = ___	11 X 7 = ___	2 X 9 = ___
8 X 4 = ___	11 X 6 = ___	9 X 7 = ___	5 X 9 = ___
12 X 4 = ___	9 X 6 = ___	2 X 7 = ___	1 X 9 = ___
6 X 4 = ___	4 X 6 = ___	3 X 7 = ___	9 X 9 = ___
1 X 4 = ___	3 X 6 = ___	7 X 7 = ___	4 X 9 = ___
10 X 4 = ___	6 X 6 = ___	5 X 7 = ___	12 X 9 = ___
3 X 4 = ___	8 X 6 = ___	12 X 7 = ___	7 X 9 = ___
11 X 4 = ___	7 X 6 = ___	8 X 7 = ___	11 X 9 = ___

time ___ sec. ___ 4's   time ___ sec. ___ 6's   time ___ sec. ___ 7s   time ___ sec. ___

Wrapping Time _____

RAPID WRITING EXERCISE WORKSHEET

Think about the commutative partners of 8 and write all the answers you know. →

6 X 6 = ___	11 X 7 = ___	6 X 9 = ___	6 X 8 = ___
1 X 6 = ___	9 X 7 = ___	10 X 9 = ___	3 X 8 = ___
8 X 6 = ___	2 X 7 = ___	4 X 9 = ___	8 X 8 = ___
3 X 6 = ___	3 X 7 = ___	1 X 9 = ___	10 X 8 = ___
7 X 6 = ___	7 X 7 = ___	11 X 9 = ___	2 X 8 = ___
10 X 6 = ___	5 X 7 = ___	9 X 9 = ___	5 X 8 = ___
5 X 6 = ___	12 X 7 = ___	2 X 9 = ___	1 X 8 = ___
12 X 6 = ___	8 X 7 = ___	3 X 9 = ___	9 X 8 = ___
2 X 6 = ___	6 X 7 = ___	7 X 9 = ___	4 X 8 = ___
11 X 6 = ___	10 X 7 = ___	5 X 9 = ___	12 X 8 = ___
9 X 6 = ___	4 X 7 = ___	12 X 9 = ___	7 X 8 = ___
4 X 6 = ___	1 X 7 = ___	8 X 9 = ___	11 X 8 = ___

time ___ sec. ___ 6's   time ___ sec. ___ 7's   time ___ sec. ___ 9's

Wrapping Time _____

$8 \times 1 =$ _____	$1 \times 8 =$ _____
$8 \times 2 =$ _____	$2 \times 8 =$ _____
$8 \times 3 =$ _____	$3 \times 8 =$ _____
$8 \times 4 =$ _____	$4 \times 8 =$ _____
$8 \times 5 =$ _____	$5 \times 8 =$ _____
$8 \times 6 =$ _____	$6 \times 8 =$ _____
$8 \times 7 =$ _____	$7 \times 8 =$ _____
$8 \times 8 =$ _____	$8 \times 8 =$ _____
$8 \times 9 =$ _____	$9 \times 8 =$ _____
$8 \times 10 =$ _____	$10 \times 8 =$ _____
$8 \times 11 =$ _____	$11 \times 8 =$ _____
$8 \times 12 =$ _____	$12 \times 8 =$ _____

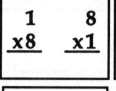

| 1 8 | 2 8 | 3 8 | 4 8 | 5 8 | 8x6=___ |
| x8 x1 | x8 x2 | x8 x3 | x8 x4 | x8 x5 | 6x8=___ |

| 8x7=___ | 8 x8 | 8x9=___ | 8x10=___ | 8x11=___ | 8x12=___ |
| 7x8=___ | | 9x8=___ | 10x8=___ | 11x8=___ | 12x8=___ |

1. Matt could see 7 rows of bricks on one side of the flagpole base. Each row had 8 bricks. Draw the bricks. Write the problem and answer.

_____

The bottom row of bricks is already drawn.

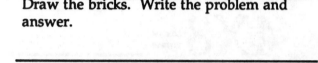

2. Marianne counted 8 tiles in a column at the school entrance. There were 7 columns. How many tiles were there? Write the problem and answer.

_____

The first column is drawn for you.

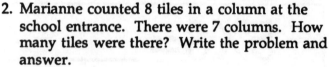

3. Ron went to the egg farm on a field trip. He noticed 8 small egg cartons that held only 6 eggs.   Draw a picture of the eggs in the cartons. Write the problem and answer.

_____

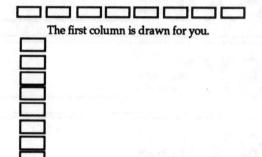

Draw the rest of the egg cartons and eggs.

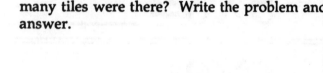

Draw the rest of the counters on the paper.

4. Rich put plastic counters on a paper. He put 8 counters in a row. There were 6 rows. Draw the counters. Write the problem and answer.

_____

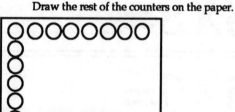

5. Write all of the 8X ___ = ___ problems and their commutative partners ___ X 8 =___.

_____ _____ _____ _____

_____ _____ _____ _____

_____ _____ _____ _____

_____ _____ _____ _____

NUMBER FAMILY FUNSHEET
____ X8 or 8X ____
SKIP COUNTING

Write the 8's skip counting numbers in the OCTAGONS 2 times.

(8) (40) (24) (96) (82) (80) (48) (32) (64) (58) (42)

Cross out the numbers that are not part of the X8 family.

(13)

(24) (56) (96) (64) (72) (80) (68) (16) (72) (88) (56)

(32)

(40) (86) (48) (16) (88) (58) (42) (18) (62) (32)

(8)

Octagons column:
8 16 24 32 40 48 56 64 72 80 88 96

Start

| 8 | 32 | 40 | 48 | 56 | 64 |
| 16 | 24 | 96 | 88 | 80 | 72 |

56	48	8
64	40	16
72	32	24
80	88	96
56	48	8
64	40	16
72	32	24
80	40	48
88	32	56
96	24	64
8	16	72
96	88	80

End

Start at the top and make a trail with your pencil through the X8 number family. Each time you get to 96, start over with 8. Can you get to the end without lifting your pencil off the paper?

Cover the skip counting numbers on the left side of the page, then write all the X8 family numbers you can remember. Two of them are done.

VISITING THE FARM   Multiply by 8

Answers

Show how you were
thinking to find the answer.

1. The farmer told _____
   that he has 8 small horses. Each one gets
   2 bales of hay a week. How many bales of
   hay do the horses get each week?

   _____

2. _____ counted 4
   horseshoes on each horse. There were 8
   horses. How many horseshoes?

   _____

3. _____ , _____
   and _____ each carried 8
   baby bottles of milk out to feed the lambs.
   If there was a lamb for each bottle of
   milk, how many lambs were there?

   _____

4. _____ carried 5 buckets
   of corn to the pigs. Each bucket had 8
   ears of corn in it. How many ears of
   corn were carried to the pigs?

   _____

5. The farmer let _____ help
   him drive the seed planting tractor. It
   planted 8 rows of seeds at a time. They
   drove the tractor across the field 7 times.
   How many rows of seeds did they plant?

   _____

6. At the chicken pen _____
   gathered 9 eggs from each nest. There
   were 8 nests. How many eggs did they
   gather?

   _____

7. It was 8 miles out to the farm and back.
   _____ figured how many
   miles it would be if they went to the farm
   8 times. How many miles would it be?

   _____

8. There were 10 cows in the milking barn.
   _____ saw that each cow
   wore a collar with 8 bells on it. How
   many bells were there on all the cows?

   _____

9. _____ fed the 8 puppies 6 dog
   biscuits each. How many biscuits for all
   the puppies?

   _____

DAY 8

CHECKING PROGRESS

- Take HERE'S WHAT I'VE LEARNED from folder. Students **write** all the answers to the problems for the #1 and #2 facts. Teacher reads the correct answers and students correct. Tell them to mark the errors and write the correct answer.

- Color all the #1 and #2 problems and answers yellow. (See the book cover.) Follow the same procedure with each of the following numbers:

    #3 and #10 problems not previously answered. Correct and color orange.

    #4 and #11 problems not previously answered. Correct and color red.

    #5 and perfect square problems not previously answered. Correct and color purple.

    Students should have answered #6 previously. Color them blue.

- Write the #7, #9, and #8 answers that have not been completed. Correct and color 7's and 9's green. Color the #8 problems pink.

    Students have now completed the Mathnique Coloring of HERE'S WHAT I'VE LEARNED. They should know all the facts.

- Have students complete the PRACTICE WRITE-ON SHEET exercise. Teacher or volunteer corrects. Group students who may be having trouble in a specific area. Plan to work with any problems that have not been mastered on DAY 9.

- The remainder of the 10 DAYS should be used for Learning Wrap-up competition, which will increase the speed of the student recall.

TODAY'S LESSON PRACTICE, and WRAP-OFFS

A WRAP-OFF is where the student wraps one board after another as fast as they can.

- Have student choose the Wrap-ups that are hardest for them FIRST. It will give them more practice in the area where they need it.

- Have students do Wrap-offs for about 20 minutes without being timed. (This gives the teacher time to go through the WRITE-ON PRACTICE SHEETS)

- Do WRAP-OFF. Time and record. GOAL: less than 5 minutes. **No one sits down.** As students finish they should record their time, check the back of the boards for correctness, then pick up a difficult board and do it again. In 5 minutes call time and have everyone sit down. Students who have not finished should do so now.

SEND HOME

- Practice Write On. Students need to practice doing the whole page without stopping. Ask parents to time how long it takes and correct. Instruct students to practice the Wrap-ups they need the most help with.

- A complete set of Learning Wrap-ups. Practice for the big WRAP-OFF on DAY 10.

- **STRESS IMPORTANCE OF BRINGING THE WRAP-UPS BACK!**

PRACTICE   WRITE-ON

Write all the answers you know, then go back and work on the ones you have to think about.
THINK commutative partners!   WRITE AS FAST AS YOU CAN!

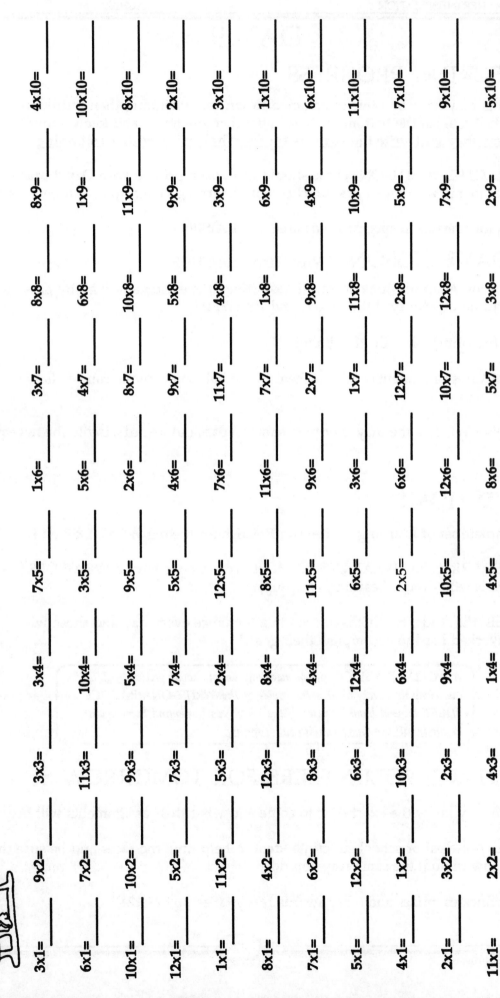

3x1= ___	9x2= ___	3x3= ___	3x4= ___	7x5= ___	1x6= ___	3x7= ___	8x8= ___	8x9= ___	4x10= ___
6x1= ___	7x2= ___	11x3= ___	10x4= ___	3x5= ___	5x6= ___	4x7= ___	6x8= ___	1x9= ___	10x10= ___
10x1= ___	10x2= ___	9x3= ___	5x4= ___	9x5= ___	2x6= ___	8x7= ___	10x8= ___	11x9= ___	8x10= ___
12x1= ___	5x2= ___	7x3= ___	7x4= ___	5x5= ___	4x6= ___	9x7= ___	5x8= ___	9x9= ___	2x10= ___
1x1= ___	11x2= ___	5x3= ___	2x4= ___	12x5= ___	7x6= ___	11x7= ___	4x8= ___	3x9= ___	3x10= ___
8x1= ___	4x2= ___	12x3= ___	8x4= ___	8x5= ___	11x6= ___	7x7= ___	1x8= ___	6x9= ___	1x10= ___
7x1= ___	6x2= ___	8x3= ___	4x4= ___	11x5= ___	9x6= ___	2x7= ___	9x8= ___	4x9= ___	6x10= ___
5x1= ___	12x2= ___	6x3= ___	12x4= ___	6x5= ___	3x6= ___	1x7= ___	11x8= ___	10x9= ___	11x10= ___
4x1= ___	1x2= ___	10x3= ___	6x4= ___	2x5= ___	6x6= ___	12x7= ___	2x8= ___	5x9= ___	7x10= ___
2x1= ___	3x2= ___	2x3= ___	9x4= ___	10x5= ___	12x6= ___	10x7= ___	12x8= ___	7x9= ___	9x10= ___
11x1= ___	2x2= ___	4x3= ___	1x4= ___	4x5= ___	8x6= ___	5x7= ___	3x8= ___	2x9= ___	5x10= ___
9x1= ___	8x2= ___	1x3= ___	11x4= ___	1x5= ___	10x6= ___	6x7= ___	7x8= ___	12x9= ___	12x10= ___

Total Writing time _____

91

DAY 9

CHECKING PROGRESS

Students may have become accustomed to writing a column, then putting the number of seconds it took, at the bottom. It is essential that you stress and **keep reminding** students that <u>**now they will write the answers for the whole page without stopping.**</u>

• Do PRACTICE WRITE-ON. These should have been printed with Day 8 practice copies. Have students who finish first practice with Learning Wrap-ups. Students correct.

• Group for practice in specific needs area.   PRAISE!

TODAY'S LESSON REVIEW and PRACTICE

• Practice for Wrap-off three or four times. (Keep the excitement high by passing out Mathnique Money, little prizes, and PRAISE!)

• Invite the principal to do the timing.

• Take pictures of students and put them in the hall so everyone can see how successful they are.

• Get publicity! Ask the mayor or merchants to pass out awards, invite the newspaper!

SEND HOME

• A complete set of Learning Wrap-ups. Practice for the big WRAP-OFF on DAY 10.

• STRESS BRINGING WRAP-UPS BACK !!! You cannot have a WRAP-OFF if anyone is left without Learning Wrap-ups.

• PRAISE the students who have been in attendance every day and those who have worked hard to accomplish their goals!

> *A WORD OF ADVICE: If you have any doubts about getting <u>ALL</u> of the Wrap-ups back, print extra copies of the WRITE-ON PRACTICE SHEET to send home in place of the Wrap-ups.   Remind them again to <u>write all the answers without stopping.</u>*

CONTACT VOLUNTEERS FOR TOMORROW TC

• Be sure they know the exact time to come and what their assignments will be.

• Ask the principal or school superintendent to help time the races and be sure they know how to do it the same way you do.

• Get certificates, prizes and other awards ready. (See pp 123-124)

DAY 10

1. WRITE-ON

Pass out OFFICIAL WRITE-ON. Tell students to complete it as fast as they can. Older students should take less than five minutes. Younger ones should be allowed at least six minutes. Have papers corrected immediately by volunteers.

2. WRAP-OFF COMPETITION

You may want to put students in groups of six. Small group competition makes it easier to get an accurate completion time for each student.   Everyone who finishes in five minutes or less gets an award. Add 10 seconds to the final time for each error.

3. TOTAL CLASSROOM WRAP-OFF

First students raise their hands as they finish. Have someone write the names of students and order they finish. Check their Wrap-ups to be certain they have been done correctly.   If there are errors, add 10 seconds to the final time for each one.

4. PASS OUT AWARDS FOR

- FIRST through ___You decide!___ PLACE
- MOST TIME SPENT ON HOMEWORK
  Honor everyone who did what was required or more.
- BEING IN ATTENDANCE EVERY DAY
- FASTEST WRAPPERS
- ALL WHO COMPLETE THEIR WRITE-ON SHEET
  IN GOAL TIME
- FASTEST WRITERS
- ANY OTHER THING YOU CAN THINK OF

5. PLAY GAMES from the Extended Activities Section 4 for the party.

6. SERVE COOKIES, hopefully made by the lunchroom cooks, using the recipe on p. 98 section 4.

Hurrah !!!!!

IN THE WEEKS TO FOLLOW, play CrossPath (all formats), HiLo and Bingo. Keep students' minds sharp and their self-image high.   Give them plenty of multiplication problems to apply their new knowledge to.

NAME _____ # _____

OFFICIAL WRITE-ON

Write as fast as you can!

6x1= ___	11x2= ___	9x3= ___	11x4= ___	2x5= ___	4x6= ___	1x7= ___	5x8= ___	8x9= ___	10x10= ___
1x1= ___	3x2= ___	11x3= ___	9x4= ___	3x5= ___	6x6= ___	8x7= ___	12x8= ___	12x9= ___	4x10= ___
8x1= ___	9x2= ___	2x3= ___	3x4= ___	7x5= ___	5x6= ___	3x7= ___	8x8= ___	6x9= ___	1x10= ___
3x1= ___	7x2= ___	3x3= ___	4x4= ___	5x5= ___	2x6= ___	4x7= ___	6x8= ___	1x9= ___	2x10= ___
7x1= ___	4x2= ___	7x3= ___	6x4= ___	12x5= ___	7x6= ___	9x7= ___	10x8= ___	11x9= ___	3x10= ___
10x1= ___	10x2= ___	5x3= ___	5x4= ___	8x5= ___	11x6= ___	11x7= ___	4x8= ___	9x9= ___	8x10= ___
5x1= ___	5x2= ___	12x3= ___	7x4= ___	6x5= ___	9x6= ___	2x7= ___	1x8= ___	3x9= ___	6x10= ___
12x1= ___	6x2= ___	8x3= ___	2x4= ___	10x5= ___	3x6= ___	12x7= ___	9x8= ___	4x9= ___	11x10= ___
2x1= ___	12x2= ___	6x3= ___	8x4= ___	4x5= ___	8x6= ___	5x7= ___	11x8= ___	10x9= ___	7x10= ___
11x1= ___	1x2= ___	10x3= ___	12x4= ___	1x5= ___	12x6= ___	6x7= ___	2x8= ___	5x9= ___	5x10= ___
9x1= ___	2x2= ___	4x3= ___	10x4= ___	11x5= ___	1x6= ___	7x7= ___	3x8= ___	7x9= ___	12x10= ___
4x1= ___	8x2= ___	1x3= ___	1x4= ___	9x5= ___	10x6= ___	10x7= ___	7x8= ___	2x9= ___	9x10= ___

Total Time _____ minutes _____ seconds

Number Correct _____

94

SECTION FOUR

FOLLOW-UP AND EXTENDED ACTIVITIES AND TEACHER TOOLS

FOLLOW-UP AND EXTENDED ACTIVITIES

PRACTICE and FOLLOW-UP !

Using multiplication facts is much like playing the piano. The skill will always be there once you've learned it, but it does get rusty without practice. Keep your students practicing. Frequent Wrap-offs keep kids sharp. If you have a competition of one kind or another and give out rewards on a weekly basis, students will practice in their spare time. (Rewards can be certificates, praise, trips to the principal for compliments, candy, just about anything.)

Tape a big red **X** on your classroom door   each week as your students review and pass a weekly   workout. Have all the students that passed, sign the

APPLICATION !

Rapid fire repeating of math facts is of little use unless it is applied to life. The following ideas can keep kids excited about what is happening in mathematics.

STORY PROBLEMS !

Have your students write four or five story problems for each set of multiplication facts. Simple problem solving always involves two items of a certain quantity and a question about those items.

MATH PUZZLES !

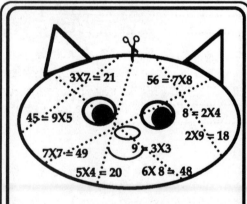

$3 \times 7 = 21$ $56 = 7 \times 8$

$45 = 9 \times 5$ $8 = 2 \times 4$

$2 \times 9 = 18$

$7 \times 7 = 49$ $9 = 3 \times 3$

$5 \times 4 = 20$ $6 \times 8 = 48$

Simple coloring book pictures can be used. With a pencil and ruler the student can draw several straight dotted lines on the picture. Then write multiplication problems on one side of the dotted line and the answer on the other side or write the problem so it is divided in the middle. Glue picture on posterboard and cut on the dotted lines. Keep pieces in a zipper type plastic bag.

Do some **BIGGY BONUS** problems.

2 x 3 = ____ x 2 =____ - 7= ____+ 4 = ___ x1=____- 5 =____ x 9 = ___ + 3=____

BIGGY BONUS !

Have students write Biggy Bonus problems. Work the problems as they are created. Then have students copy the problems, eliminating all answers following the = sign. Turn both copies in to the teacher.

MAKE A BOOK !

Make copies of all the student produced math lessons. Put in a book and give to your library. Make book copies for students to take home and work on during the summer.

VARIETY !

Cut a Rapid Writing sheet into pieces with 4 problems each piece. Glue them on another piece of paper at all different angles--upside down, sideways, etc. Make a copy for each student. Have races to complete a page. It takes the boredom out of plain old drill. Do this for a "Weekly Workout" with an official Write-On test.

PATTERNS !

GET FRIENDLY WITH 12!

Use the 12's sheet (p.101) to look for patterns. Make up questions that other students can answer. How many more multiples of 12 can be written on one sheet of paper?

Look at 12 from a different point of view. How many problems can you write with the answer of 12?

Consider addition and subtraction problems also.

ART !

Have students write one story problem on a card. Illustrate the problem on another card. Play the memory matching game. **WRITE!**

5 X 8 = 40

Blank cards are included in this book (p. 100).

Have students create their own games by writing problems on the cards.

Have a "Dozens Day". Bring or name everything that comes in dozens. Two 1/2 dozens can qualify.

COOKING !

Grandpa M's BIG Chocolate Chip Cookies

Whip until fluffy
10 eggs
4 cups shortning
6 cups sugar
5 teaspoons vanilla
1 teaspoon maple flavoring

STIR IN with a knife
8 cups oatmeal
7 cups flour
4 teaspoons soda
3 cups walnut pieces
2 packages chocolate chips

Use 1 heaping tablespoon of dough for each cookie.

BAKE at 375 degrees for 20 -25 min.

This batch makes 9 dozen BIG cookies. How much of each ingredient would be needed if you made 6 batches? What if you made 7 or 8 batches?

(Believe it or not, this is a real recipe.)

How many cookies would it take for every student in the school to have one?

Choose how many batches of cookies you would bake and multiply each of the ingredients.

Ask the lunchroom cooks to make some of these cookies for you.

CROSSPATH Product Game

Each student has a copy of the game board. Teacher calls out two factors. Student covers only **one** product that matches the 2 factors called. First player to make a path from top to bottom or side to side wins. The covered squares must touch at top, bottom or side. Touching diagonal points do not count.

VISUAL DISCRIMINATION

THE COUNTING CARDS, 6 chicks (p. 66), 7 pups (p. 81) and 8 cars (p. 89), can be used for visual discrimination activities. Each card is slightly different. Two sets printed on different colors of card stock make a great matching game.

CROSSPATH VARIATIONS

CROSSPATH Factor Game. (p.103) Teacher calls a product. Students write the product (answer) called under all the problems it belongs to. First player with a path of answered problems is the winner.

CROSSPATH Blackout. (p.103) Students cover all factors that match the product called. This could entail as many as 6 sets of factors being coverd with just one product.

Play with 2 or 3 players, each with their own color playing pieces or crayon. Print both factor and product boards on heavy paper. Cut product board into individual numbers. Turn numbers face down. Players take turns drawing and covering their choice of co-ordinating factors. (Only one square can be covered in a turn. This requires a great deal of strategy on the part of the players.)

Write the product (p.104) in the correct place when the factors are called.

MORE GAMES !

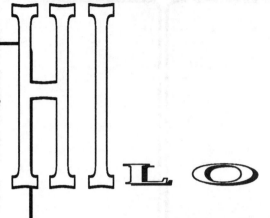

HI-LO is played similar to Bingo (p. 105)
Teacher calls out 2 factors (numbers that can be multiplied). The player chooses one of the correct answers or its commutative partner. Only one answer can be marked. When a player has marked seven products (numbers) in a row, up, down, straight across or diagonally, "HI-LO" is called. That player is the winner.

Variation: Two students take turns using different color crayons to mark one product each time the factors are called. Winner is student who is first to get seven in a row.

BINGO keeps all players thinking at the same time. Bingo cards and the check-off sheet are found on pages 116-121.

Playing pieces can be colored macaroni shells, or other shapes, buttons, colored squares of paper, etc.
If the game boards do not need to be reused, use crayons and color the squares.

TRANSITION TO DIVISION

FACT FAMILIES (pp 111-120)
Working with the number families and the inverse operation will help students get ready for division. Some teachers prefer to do these pages as they proceed through the multiplication facts. Others have found it a help before going into the division section of their text books. Students may enjoy working with the ARRAY PATTERN and a specific number of manipulatives. 12 is a good number to develop the concept with. First do 3X4 = 12, 4X3 = 12, then 12 ÷ 3 = 4 and 12÷ 4 =3. 2X6, 12÷ 6, etc.

WHICH MEMBER OF THE NUMBER FAMILY IS LOST ?

Who's missing? (p.120)
Teacher calls for a family at a certain address. (example: 6 a, 5 c, 2 h, etc.) Students find the numbered box and call out or write down the missing number.

FACTOR / WORD SEARCH

Find the factors (p. 121) in the hidden number words and look for another message.

TC BLANK CARDS Create a game.   Draw pictures and/or write problems first.   Have teacher check them, then cut out cards.   NAME _____ #

Students Own Problem Solving Project
Print cards on heavy paper. Each student writes 10 problems, 1 per card. Several students put cards together. Stack cards face down. Take turns, turn up a card, read aloud and give the correct answer. If an error is made, the card goes to the next person. Students keep the cards they answer correctly. Person with the most cards at the end wins.

Matching Game
Print two copies per student on heavy paper. Students write a story problem on one card. Illustrate it on another. Then two students play a matching game. Turn all cards face down. Student draws two cards. If they are a match, cards are kept. If not, cards are put back. Winner is the person with the most matches at the end of the game.

100

© Learning Wrap-ups Inc. 1996

MAKING FRIENDS WITH 12's FUNSHEET

Can you see any patterns in these numbers which are multiples of 12?
List them at the bottom.

Can you point to the product of 12 x 30?   60?   75?   90? How about 12 X 17?

12 x 0 =	0	240	480	720	960
	12	252	492	732	972
	24	264	504	744	984
	36	276	516	756	996
	48	288	528	768	1008
12 x 5 =	60	300	540	780	1020
	72	312	552	792	1032
	84	324	564	804	1044
	96	336	576	816	1056
	108	348	588	828	1068
12 x 10 =	120	360	600	840	1080
	132	372	612	852	1092
	144	384	624	864	1104
	156	396	636	876	1116
	168	408	648	888	1128
12 x 15 =	180	420	660	900	1140
	192	432	672	912	1152
	204	444	684	924	1164
	216	456	696	936	1176
	228	468	708	948	1188

Is 847 a multiple of 12? Why?

How many multiples of 12 start with 800? Is that more, less, or the same as multiples that start with 500?

Is there a pattern for the numbers in the 1's column? How about the 10's column? Is there a pattern? What?

396 is the product of 12 X what?

Does 12 X a number with a 0 in the 1's column always have a product with a 0 in the 1's column? Why or why not?

1200=12x100

_____

_____

_____

_____

CROSSPATH (PRODUCT) GAME BOARD

Teacher calls two factors. Student chooses one of the correct spaces and covers the product (answer) with a marker.   First student to cover a path of numbers from top to bottom or side to side wins. Variation: Two or three students take turns covering the product. This requires careful strategy to become a winner. Add excitement by calling the factors in rapid succession.

1	2	3	4	5	6	7	8	9	10	11	12
2	4	6	8	10	12	14	16	18	20	22	24
3	6	9	12	15	18	21	24	27	30	33	36
4	8	12	16	20	24	28	32	36	40	44	48
5	10	15	20	25	30	35	40	45	50	55	60
6	12	18	24	30	36	42	48	54	60	66	72
7	14	21	28	35	42	49	56	63	70	77	84
8	16	24	32	40	48	56	64	72	80	88	96
9	18	27	36	45	54	63	72	81	90	99	108
10	20	30	40	50	60	70	80	90	100	110	120
11	22	33	44	55	66	77	88	99	110	121	132
12	24	36	48	60	72	84	96	108	120	132	144

CROSSPATH (FACTOR) GAME BOARD

Teacher calls out a product (answer). Player covers all the factor combinations (numbers) belonging to the product. First player to cover a path from top to bottom or side to side wins. Variations: • Two or three players take turns covering only one set of factors for each product. • Play "Blackout" by covering all factor combinations for each product called.

1 x1	1 x2	1 x3	1 x4	1 x5	1 x6	1 x7	1 x8	1 x9	1 x10	1 x11	1 x12
2 x1	2 x2	2 x3	2 x4	2 x5	2 x6	2 x7	2 x8	2 x9	2 x10	2 x11	2 x12
3 x1	3 x2	3 x3	3 x4	3 x5	3 x6	3 x7	3 x8	3 x9	3 x10	3 x11	3 x12
4 x1	4 x2	4 x3	4 x4	4 x5	4 x6	4 x7	4 x8	4 x9	4 x10	4 x11	4 x12
5 x1	5 x2	5 x3	5 x4	5 x5	5 x6	5 x7	5 x8	5 x9	5 x10	5 x11	5 x12
6 x1	6 x2	6 x3	6 x4	6 x5	6 x6	6 x7	6 x8	6 x9	6 x10	6 x11	6 x12
7 x1	7 x2	7 x3	7 x4	7 x5	7 x6	7 x7	7 x8	7 x9	7 x10	7 x11	7 x12
8 x1	8 x2	8 x3	8 x4	8 x5	8 x6	8 x7	8 x8	8 x9	8 x10	8 x11	8 x12
9 x1	9 x2	9 x3	9 x4	9 x5	9 x6	9 x7	9 x8	9 x9	9 x10	9 x11	9 x12
10 x1	10 x2	10 x3	10 x4	10 x5	10 x6	10 x7	10 x8	10 x9	10 x10	10 x11	10 x12
11 x1	11 x2	11 x3	11 x4	11 x5	11 x6	11 x7	11 x8	11 x9	11 x10	11 x11	11 x12
12 x1	12 x2	12 x3	12 x4	12 x5	12 x6	12 x7	12 x8	12 x9	12 x10	12 x11	12 x12

CROSSPATH (GRAPH) GAME BOARD

Teacher calls product or factors. Student chooses <u>one</u> of the correct spaces and writes in the product. First student to get a path of numbers from top to bottom or side to side wins. **Variations:** • Student fills in <u>all</u> the spaces where the product (number) should be. The fastest thinker and writer will win. • Two students take turns filling in the proper squares. This requires careful strategy to become a winner.

	1	2	3	4	5	6	7	8	9	10	11	12
1												
2												
3												
4												
5												
6												
7												
8												
9												
10												
11												
12												

TC EXTENDED ACTIVITY   HI-LO GAME BOARD

Play similar to Bingo. Teacher calls out 2 factors. Player chooses the correct answer, or it's commutative partner, but can cover only one space at a time. The winner is the first person to have 7 numbers in a row, diagonally, up and down or straight across.

1	2	3	4	5	6	7	8	9	10	11	12
1	14	18	4	35	12	28	32	9	40	110	FREE
8	2	6	24	5	42	49	FREE	63	30	66	12
4	20	21	36	15	66	56	8	99	90	132	60
11	4	FREE	8	40	6	35	40	18	60	11	144
FREE	16	9	44	10	36	21	72	90	110	77	48
2	8	3	20	55	48	FREE	16	27	100	44	108
7	12	24	28	30	18	42	80	FREE	80	88	132
9	24	15	FREE	45	54	77	64	81	50	121	72
10	6	36	12	60	FREE	14	24	36	20	33	96
3	22	27	32	50	30	84	96	54	70	FREE	36
6	FREE	33	48	20	72	63	56	108	120	99	120
5	10	12	40	FREE	60	7	88	72	10	22	84
12	18	30	16	25	24	70	48	45	FREE	55	24

H I - L O

BINGO with the 4's, 6,'s, 7's, 8's, and 9's

SUGGESTIONS

- Play with preprinted cards or have students fill in the blank Bingo Cards.
- Play regular rules or have students also look for the commutative partner.
- Fill blank cards with the factors. Teacher calls out the product.
- This list can be used many times by marking with colored pencils or by marking squares as numbers are called out.
  Example: This square has been used for five games. ⊠

4	6	7	8	9
4x1=4	6x1=6	7x1=7	8x1=8	9x1=9
4X2=8	6X2=12	7X2=14	8X2=16	9X2=18
4X3=12	6X3=18	7X3=21	8X3=24	9X3=27
4X4=16	6X4=24	7X4=28	8X4=32	9X4=36
4X5=20	6X5=30	7X5=35	8X5=40	9X5=45
4X6=24	6X6=36	7X6=42	8X6=48	9X6=54
4X7=28	6X7=42	7X7=49	8X7=56	9X7=63
4X8=32	6X8=48	7X8=56	8X8=64	9X8=72
4X9=36	6X9=54	7X9=63	8X9=72	9X9=81
4X10=40	6X10=60	7X10=70	8X10=80	9X10=90
4X11=44	6X11=66	7X11=77	8X11=88	9X11=99
4X12=48	6X12=72	7X12=84	8X12=96	9X12=108

4	6	7	8	9
4x1=4	6x1=6	7x1=7	8x1=8	9x1=9
4X2=8	6X2=12	7X2=14	8X2=16	9X2=18
4X3=12	6X3=18	7X3=21	8X3=24	9X3=27
4X4=16	6X4=24	7X4=28	8X4=32	9X4=36
4X5=20	6X5=30	7X5=35	8X5=40	9X5=45
4X6=24	6X6=36	7X6=42	8X6=48	9X6=54
4X7=28	6X7=42	7X7=49	8X7=56	9X7=63
4X8=32	6X8=48	7X8=56	8X8=64	9X8=72
4X9=36	6X9=54	7X9=63	8X9=72	9X9=81
4X10=40	6X10=60	7X10=70	8X10=80	9X10=90
4X11=44	6X11=66	7X11=77	8X11=88	9X11=99
4X12=48	6X12=72	7X12=84	8X12=96	9X12=108

Choose numbers from each column to fill in the bingo cards. Use every number at least once. As the teacher calls out a problem, cover the answer on your card. Don't forget to look for the commutative partner also. Play one card at a time or play all 3.

4	6	7	8	9
4	6	7	8	9
8	12	14	16	18
12	18	21	24	27
16	24	28	32	36
20	30	35	40	45
24	36	42	48	54
28	42	49	56	63
32	48	56	64	72
36	54	63	72	81
40	60	70	80	90
44	66	77	88	99
48	72	84	96	108

Card 2 header: 4 | 6 | 7 | 8 | 9

Card 3 header: 4 | 6 | 7 | 8 | 9

Card 4 header: 4 | 6 | 7 | 8 | 9

Bingo Set A

1

4X	6X	7X	8X	9X
4	36	63	8	18
32	66	28	56	81
12	6		64	45
16	54	84	40	9
24	48	42	72	54

2

4X	6X	7X	8X	9X
28	6	28	40	18
36	72	63	16	45
8	12		96	72
32	48	21	64	54
24	36	42	56	81

3

4X	6X	7X	8X	9X
36	42	56	64	63
48	24	21	96	81
44	18		16	108
28	12	35	32	72
8	72	49	48	27

4

4X	6X	7X	8X	9X
44	54	49	72	63
16	42	35	64	108
4	66		28	27
48	24	56	32	36
12	18	84	48	9

Card 6

4X	6X	7X	8X	9X
28	6	84	40	81
36	72	63	16	45
12	24		32	9
48	18	56	48	36
8	12	14	96	72

Card 8 — Bingo Set B

4X	6X	7X	8X	9X
44	54	49	72	63
16	42	35	64	108
4	66		8	27
32	48	28	64	54
24	36	63	56	81

Card 5

4X	6X	7X	8X	9X
32	36	42	96	27
8	54	63	56	81
12	6		64	72
16	18	84	40	9
24	48	49	72	54

Card 7

4X	6X	7X	8X	9X
4	66	56	16	63
36	42	21	32	81
48	24		48	108
44	12	35	8	45
28	72	49	64	18

Bingo Set C

Card 9

4X	6X	7X	8X	9X
32	66	14	96	72
8	72	63	56	81
12	6		64	36
48	18	84	40	9
24	54	49	72	54

Card 10

4X	6X	7X	8X	9X
28	36	84	40	81
36	54	63	16	45
12	24		64	72
16	18	56	32	36
44	48	49	72	63

Card 11

4X	6X	7X	8X	9X
4	12	56	16	63
36	42	21	32	81
48	66		48	108
44	48	49	8	45
28	36	35	64	18

Card 12

4X	6X	7X	8X	9X
8	6	42	96	27
16	42	35	48	108
4	24		8	27
32	12	28	64	54
24	72	63	56	81

4X	6X	7X	8X	9X
8	12	14	96	72
36	72	63	16	45
28	6		40	18
32	48	21	64	54
24	36	42	56	81

4X	6X	7X	8X	9X
44	54	49	72	63
16	42	35	64	108
4	66		8	27
12	18	84	48	9
48	24	56	32	36

Bingo Set D

4X	6X	7X	8X	9X
4	36	63	8	18
16	54	84	40	9
12	6		64	45
32	66	28	56	81
24	48	42	72	54

4X	6X	7X	8X	9X
36	42	56	64	63
48	24	21	96	81
28	12		32	72
44	18	49	16	108
8	72	35	48	27

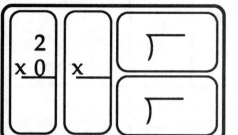

#2 FACT FAMILY

Use what you know to find the answer for each printed problem and its commutative partner, then write the inverse operation for division.

Knowing the numbers or factors and products for the FACT FAMILY make it easy. The 2X4 problem has been done for you.

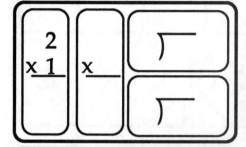

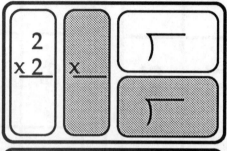

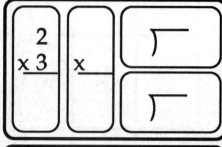

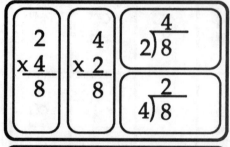

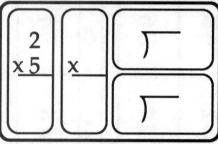

(2 x 6 box)

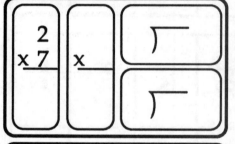

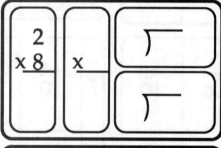

(2 x 9 box)

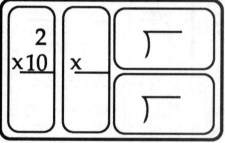

(2 x 11 box)

(2 x 12 box)

review

2 x 7 =_____	8 x 2 =_____	12 ÷ 2 =_____	6 ÷ 3 =_____
2 x 3 =_____	9 x 2 =_____	14 ÷ 2 =_____	18 ÷ 9 =_____
2 x 8 =_____	7 x 2 =_____	6 ÷ 2 =_____	16 ÷ 8 =_____
2 x 9 =_____	6 x 2 =_____	16 ÷ 2 =_____	12 ÷ 6 =_____
2 x 4 =_____	4 x 2 =_____	18 ÷ 2 =_____	14 ÷ 7 =_____
2 x 6 =_____	3 x 2 =_____	8 ÷ 2 =_____	8 ÷ 4 =_____

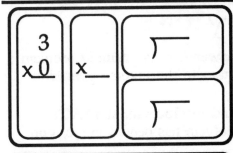

#3 FACT FAMILY

Use what you know to find the answer for each printed problem and its commutative partner, then write the inverse operation for division.

Knowing the numbers or factors and products for the FACT FAMILY make it easy. The 3X2 problem has been done for you.

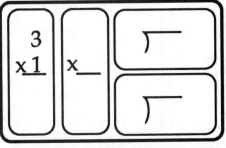 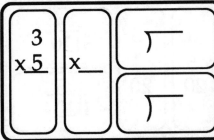 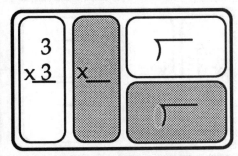

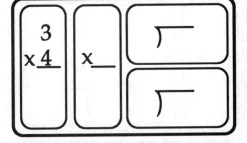 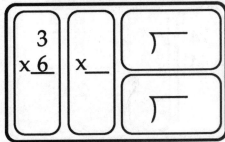 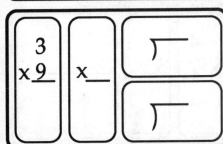

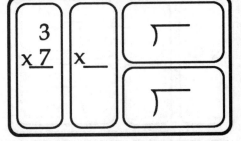

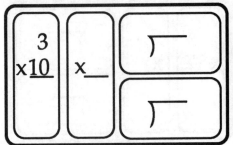 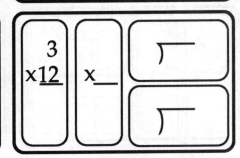

review

$3 \times 7 =$ _____	$8 \times 3 =$ _____	$21 \div 3 =$ _____	$9 \div 3 =$ _____
$3 \times 3 =$ _____	$9 \times 3 =$ _____	$9 \div 3 =$ _____	$27 \div 9 =$ _____
$3 \times 8 =$ _____	$7 \times 3 =$ _____	$24 \div 3 =$ _____	$24 \div 8 =$ _____
$3 \times 9 =$ _____	$3 \times 3 =$ _____	$9 \div 3 =$ _____	$18 \div 6 =$ _____
$3 \times 6 =$ _____	$4 \times 3 =$ _____	$18 \div 3 =$ _____	$21 \div 7 =$ _____
$3 \times 4 =$ _____	$6 \times 3 =$ _____	$12 \div 3 =$ _____	$12 \div 4 =$ _____

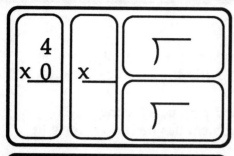

#4 FACT FAMILY

Use what you know to find the answer for each printed problem and its commutative partner, then write the inverse operation for division.

Knowing the numbers or factors and products for the FACT FAMILY make it easy. The 4X5 problem has been done for you.

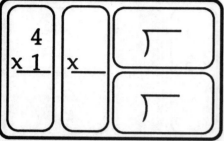

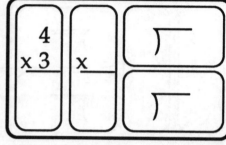

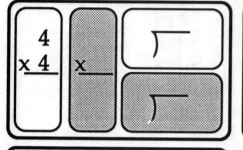

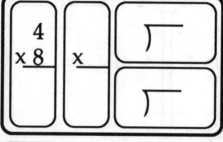

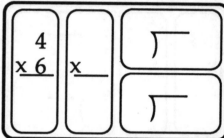

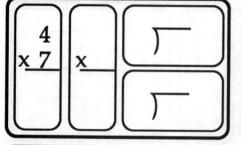

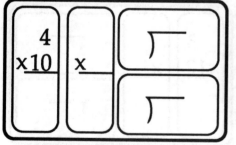

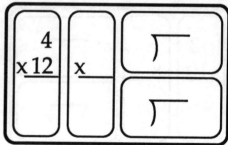

review			
4 x 7 = ____	8 x 4 = ____	12 ÷ 4 = ____	12 ÷ 3 = ____
4 x 3 = ____	9 x 4 = ____	24 ÷ 4 = ____	36 ÷ 9 = ____
4 x 8 = ____	7 x 4 = ____	36 ÷ 4 = ____	32 ÷ 8 = ____
4 x 9 = ____	6 x 4 = ____	16 ÷ 4 = ____	24 ÷ 6 = ____
4 x 4 = ____	4 x 4 = ____	28 ÷ 4 = ____	28 ÷ 7 = ____
4 x 6 = ____	3 x 4 = ____	32 ÷ 4 = ____	16 ÷ 4 = ____

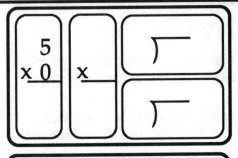

#5 FACT FAMILY

Use what you know to find the answer for each printed problem and its commutative partner, then write the inverse operation for division.

Knowing the numbers or factors and products for the FACT FAMILY make it easy. The 5X3 problem has been done for you.

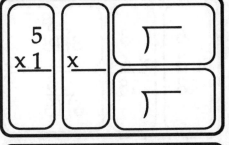

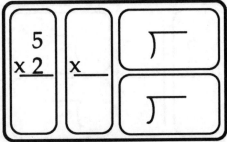

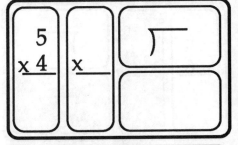

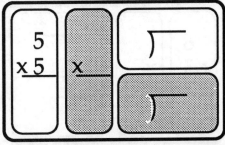

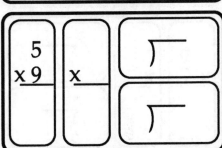

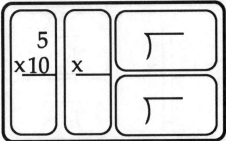

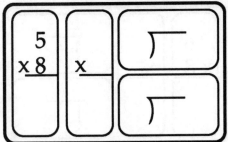

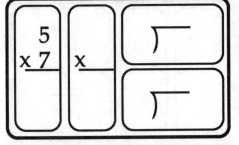

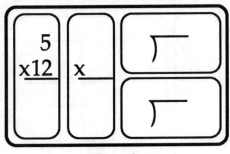

review			
5 x 7 = _____	8 x 5 = _____	15 ÷ 5 = _____	15 ÷ 3 = _____
5 x 3 = _____	9 x 5 = _____	25 ÷ 5 = _____	45 ÷ 9 = _____
5 x 8 = _____	7 x 5 = _____	40 ÷ 5 = _____	40 ÷ 8 = _____
5 x 9 = _____	6 x 5 = _____	35 ÷ 5 = _____	30 ÷ 6 = _____
5 x 4 = _____	4 x 5 = _____	45 ÷ 5 = _____	35 ÷ 7 = _____
5 x 6 = _____	3 x 5 = _____	20 ÷ 5 = _____	20 ÷ 4 = _____

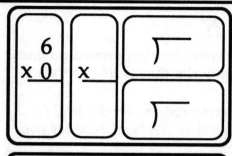

#6 FACT FAMILY

Use what you know to find the answer for each printed problem and its commutative partner, then write the inverse operation for division.

Knowing the numbers or factors and products for the FACT FAMILY make it easy. The 6X3 problem has been done for you.

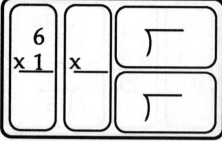

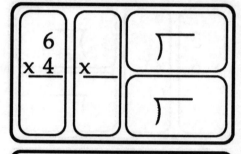

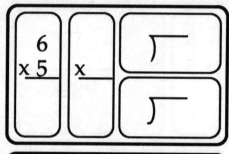

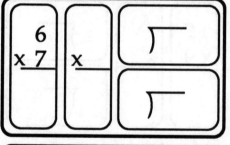

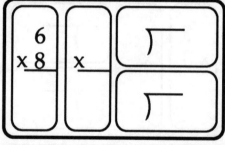

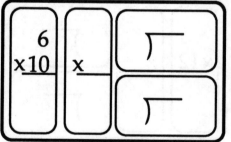

review

6 × 7 = ____	8 × 6 = ____	12 ÷ 6 = ____	48 ÷ 8 = ____
6 × 3 = ____	9 × 6 = ____	36 ÷ 6 = ____	18 ÷ 3 = ____
6 × 8 = ____	7 × 6 = ____	24 ÷ 6 = ____	42 ÷ 7 = ____
6 × 9 = ____	6 × 6 = ____	42 ÷ 6 = ____	54 ÷ 9 = ____
6 × 4 = ____	4 × 6 = ____	18 ÷ 6 = ____	36 ÷ 6 = ____
6 × 6 = ____	3 × 6 = ____	54 ÷ 6 = ____	24 ÷ 4 = ____

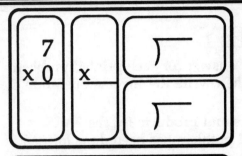

#7 FACT FAMILY

Use what you know to find the answer for each printed problem and its commutative partner, then write the inverse operation for division.

Knowing the numbers or factors and products for the FACT FAMILY make it easy. The 7X5 problem has been done for you.

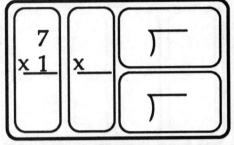

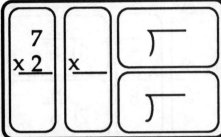

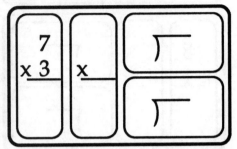

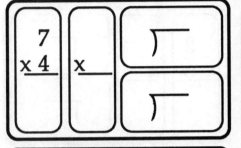

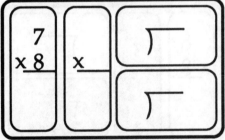

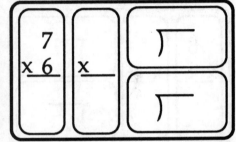

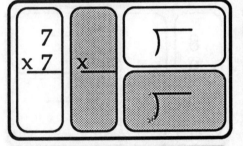

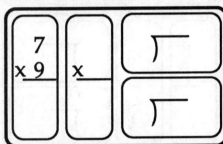

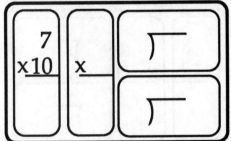

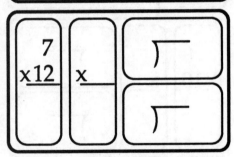

review			
7 x 7 = ____	8 x 7 = ____	63 ÷ 7 = ____	56 ÷ 8 = ____
7 x 3 = ____	9 x 7 = ____	42 ÷ 7 = ____	21 ÷ 3 = ____
7 x 8 = ____	7 x 7 = ____	56 ÷ 7 = ____	49 ÷ 7 = ____
7 x 9 = ____	6 x 7 = ____	49 ÷ 7 = ____	63 ÷ 9 = ____
7 x 4 = ____	4 x 7 = ____	21 ÷ 7 = ____	42 ÷ 6 = ____
7 x 6 = ____	3 x 7 = ____	28 ÷ 7 = ____	28 ÷ 4 = ____

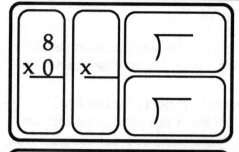

#8 FACT FAMILY

Use what you know to find the answer for each printed problem and its commutative partner, then write the inverse operation for division.

Knowing the numbers or factors and products for the FACT FAMILY make it easy. The 8X2 problem has been done for you.

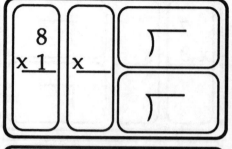

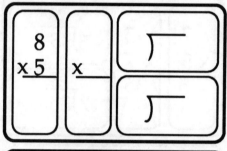

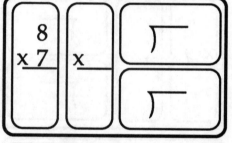

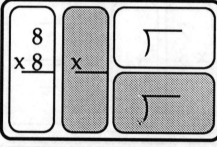

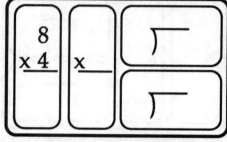

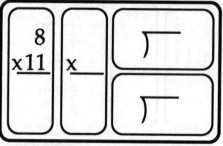

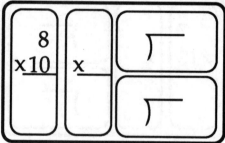

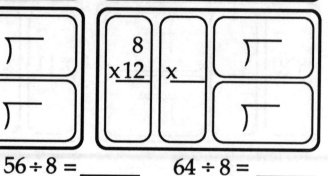

review			
8 × 7 = ___	8 × 8 = ___	56 ÷ 8 = ___	64 ÷ 8 = ___
8 × 3 = ___	9 × 8 = ___	64 ÷ 8 = ___	24 ÷ 3 = ___
8 × 8 = ___	7 × 8 = ___	72 ÷ 8 = ___	56 ÷ 7 = ___
8 × 9 = ___	6 × 8 = ___	24 ÷ 8 = ___	72 ÷ 9 = ___
8 × 4 = ___	4 × 8 = ___	32 ÷ 8 = ___	48 ÷ 6 = ___
8 × 6 = ___	3 × 8 = ___	48 ÷ 8 = ___	32 ÷ 4 = ___

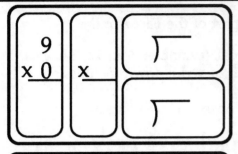

#9 FACT FAMILY

Use what you know to find the answer for each printed problem and its commutative partner, then write the inverse operation for division.

Knowing the numbers or factors and products for the FACT FAMILY make it easy. The 9X3 problem has been done for you.

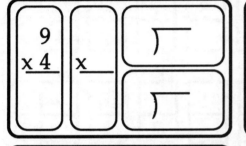

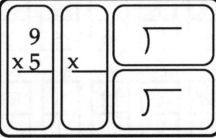

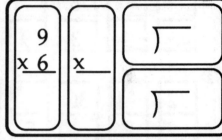

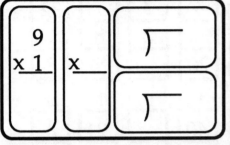

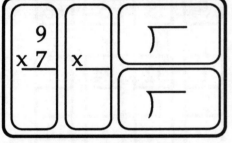

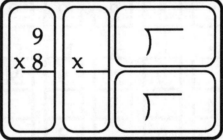

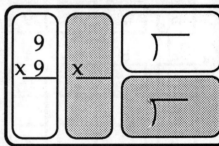

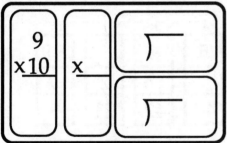

review

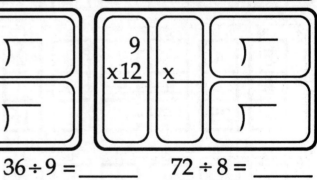

$9 \times 7 =$ ____	$8 \times 9 =$ ____	$36 \div 9 =$ ____	$72 \div 8 =$ ____
$9 \times 3 =$ ____	$9 \times 9 =$ ____	$72 \div 9 =$ ____	$27 \div 3 =$ ____
$9 \times 8 =$ ____	$7 \times 9 =$ ____	$63 \div 9 =$ ____	$63 \div 7 =$ ____
$9 \times 9 =$ ____	$6 \times 9 =$ ____	$54 \div 9 =$ ____	$81 \div 9 =$ ____
$9 \times 4 =$ ____	$4 \times 9 =$ ____	$81 \div 9 =$ ____	$54 \div 6 =$ ____
$9 \times 6 =$ ____	$3 \times 9 =$ ____	$27 \div 9 =$ ____	$36 \div 4 =$ ____

WHO'S MISSING? Think about numbers as part of the Fact Family.

- Teacher calls for a family at a certain address. (example: 6 a, 5 c, 2 h, etc.) Students find the number boxes, and call out or write down the missing number.
- Fill in the missing number in each of the boxes. Time yourself.

1.

a	b	c	d	e	f	g	h	i	j	k	l	m
2	4		5	5	2	4	2		4	3		32
	8	6		7		4		9		3	8	
2		24	30		6		8	36	20		24	8

2.

a	b	c	d	e	f	g	h	i	j	k	l	m
25	28		18	81	16	18		48	40		21	2
		9		9	8		9		8	6		
5	4	3	9			6	6	8		7	3	4

3.

a	b	c	d	e	f	g	h	i	j	k	l	m
12		36		64		12		63	15		10	7
	7	5	5		9		1			7		
6	7	6	9	8	8	3	2	9	5	8	5	63

4.

a	b	c	d	e	f	g	h	i	j	k	l	m
	35	6	16		36	20	9	24	1	32	24	2
5		2		4		5		3	2	4		
6	5		4	2	9		3				4	14

5.

a	b	c	d	e	f	g	h	i	j	k	l	m
5	4		2	9	2	3			5	6		81
5		9	9	9		6	9	8			7	
	28	27			16		54	48	40	42	21	9

6.

a	b	c	d	e	f	g	h	i	j	k	l	m
6	5	8	8	3	2	7	2		3		2	
6		8		4	1	9		7	5	8	6	8
	45		72				10	49		56		6

TIME _____ Minutes _____ Seconds

FACTOR SEARCH FUNSHEET

- Circle each number word and write the number above it. This number will be the product.
  A hyphen is used between words that are all part of one number.

- Place a mark in the factor box every time that factor is found in the product.
  Use only factors 1 thru 9. The first line is done for you.

  Example: 2=2X1 We have put marks in the 2 and the 1 boxes.
  9=3X3 and 9X1 We put marks in boxes 3, 9, 1.
  8=2X4 and 1X8 We put marks in boxes 2,4,1,8.
  14=2X7 We marked boxes 2 and 7.
  5=5X1 Boxes 1 and 5 are marked.

- Write the number of marks in each factor box beside the box number.
  Compare the numbers with those of your friends. Did you get
  them all? Did you find anything else? Check to see if others did.

FACTOR BOXES

1	ΙΙΙΙ
2	ΙΙΙ
3	Ι
4	Ι
5	Ι
6	
7	Ι
8	Ι
9	Ι
10	

Puzzle grid:

```
 2        9        8        14       5
(two)what(nine)elseca(eighty)ou(fourteen)see(five)

twenty-fivecanyoutwentyseetwenty-oneafour

messagesixintwenty-eighttheseforty-twolines

sixteenyoufiftyareeighty-oneobservantsixty

puzzlestwelvecansixbesevenone-hundredfun

forty-nineifthirty-fiveyoueighteenlookninety

hardthirty-sixyoufifteencanthirtyfortyfindten

twenty-fourmostsixty-fouroftwenty-seventhe

thirty-twonumbersforty-eightonfifty-fourthe

fifty-sixtimesforty-fivetableseventy-twochart

wehopeeightyousixty-threehavehadlotsoffun
```

**If students are numbered, this record sheet can be duplicated and used   purpose _____
for many purposes. There is no need to rewrite students' names.**

goal	1	2	3	4	5	6	7	8	9	10
1										
2										
3										
4										
5										
6										
7										
8										
9										
10										
11										
12										
13										
14										
15										
16										
17										
18										
19										
20										
21										
22										
23										
24										
25										
26										
27										
28										
29										
30										
31										
32										
33										
34										
35										
36										
37										
38										
39										
40										
41										
42										

Use for setting and meeting goals, once � twice ⊠, writing in an indivduals goal, [21] checking
Wrap-ups out � and in ⊠ recording scores on Rapid Writer Exercises, homework slips turned in,
number of Mathnique Money slips received, attendance for 10 days, etc.

CERTIFICATE

of

RECOGNITION

is awarded to

for

Outstanding Effort

in learning the

MULTIPLICATION FACTS

date

teacher

principal

date

is presented this

Award

in recognition of

teacher

principal

We hereby recognize

_____
student

as a

MATH CHAMPION

who

set a goal to

LEARN THE MULTIPLICATION FACTS

in **10 DAYS**

and has

Successfully Met The Goal!

_____
teacher

_____
principal

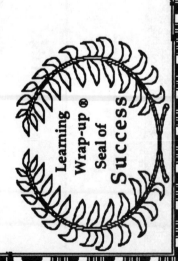

Learning
Wrap-up ®
Seal of
Success

ANSWERS for DAILY RAPID WRITING EXERCISES

The specific Rapid Writing Exercise sheets can be found with daily lesson materials. They can be used on a daily basis to assess the progress students have made each evening with their practice at home. These sheets will tell you whenever you need an extra day or more materials for a set of facts.

Day 2 Rapid Writing Exercise (p. 35) Answer Sheet

11	2	60
9	18	10
2	6	80
3	8	30
7	20	70
5	10	100
12	14	50
8	4	120
6	16	20
10	24	110
4	12	90
1	22	40

Follow-up Rapid Writing Exercise (p. 35) Answer Sheet

48	42	88
24	7	72
60	56	24
12	21	32
6	49	80
42	70	40
30	35	56
54	84	16
66	14	64
18	77	96
72	63	48
36	28	8

Day 3 Rapid Writing Exercise (p. 44) Answer Sheet

6	10	33	40
1	12	27	60
8	24	6	50
3	2	9	70
7	4	21	20
10	16	15	110
5	22	36	100
12	6	24	30
2	18	18	80
11	14	30	120
9	8	12	90
4	20	3	10

Day 4 Rapid Writing Exercise (p. 44) Answer Sheet

33	44	66
27	36	11
6	12	88
9	16	33
21	40	77
15	20	110
36	28	55
24	8	132
18	32	22
30	48	121
12	24	99
3	4	44

Day 5 Rapid Writing Exercise (p. 63) Answer Sheet

33	44	30	144
27	36	5	49
6	12	40	64
9	16	15	36
21	24	35	100
15	20	50	16
36	28	25	1
24	8	60	25
18	32	10	121
30	48	50	81
12	40	45	4
3	4	20	9

Day 6 Rapid Writing Exercise (p. 63) Answer Sheet

44	48	60	77
36	24	30	63
12	60	80	14
16	12	100	21
40	6	20	49
20	42	50	35
28	30	10	84
8	54	90	56
32	66	40	42
48	18	120	70
24	72	70	28
4	36	110	7

Day 7 Rapid Writing Exercise (p. 84) Answer Sheet

16	60	42	54
36	30	70	27
20	72	28	72
28	6	7	90
8	12	77	18
32	66	63	45
48	54	14	9
24	24	21	81
4	18	49	36
40	36	35	108
12	48	84	63
44	42	56	99

Day 8 Rapid Writing Exercise (p. 84) Answer Sheet

36	77	54	48
6	63	90	24
48	14	36	64
18	21	9	80
42	49	99	16
60	35	81	40
30	84	18	8
72	56	27	72
12	42	63	32
66	70	45	96
54	28	108	56
24	7	72	88

PRACTICE: WRITE-ON
Answer Key

Copy, cut out area indicated and
lay answer key over completed
Practice Write-on (p. 91) to correct.

cut out
40 · 100 · 80 · 20 · 30 · 10 · 60 · 110 · 70 · 90 · 50 · 120

cut out
72 · 9 · 99 · 81 · 27 · 54 · 36 · 90 · 45 · 63 · 18 · 108

cut out
64 · 48 · 80 · 40 · 32 · 8 · 72 · 88 · 16 · 96 · 24 · 56

cut out
21 · 28 · 56 · 63 · 77 · 49 · 14 · 7 · 84 · 70 · 35 · 42

cut out
6 · 30 · 12 · 24 · 42 · 66 · 54 · 18 · 36 · 72 · 48 · 60

cut out
35 · 15 · 45 · 25 · 60 · 40 · 55 · 30 · 10 · 50 · 20 · 5

cut out
12 · 40 · 20 · 28 · 8 · 32 · 16 · 48 · 24 · 36 · 4 · 44

cut out
9 · 33 · 27 · 21 · 15 · 36 · 24 · 18 · 30 · 6 · 12 · 3

cut out
18 · 14 · 20 · 10 · 22 · 8 · 12 · 24 · 2 · 6 · 4 · 16

cut out
3 · 6 · 10 · 12 · 1 · 8 · 7 · 5 · 4 · 2 · 11 · 9

126

TEACHER   Rm #

ANSWER KEY
for Official Write-on page 94

OFFICIAL WRITE-ON

Copy, cut out, lay sheet over Official Write-on to correct.

cut out	cut out	cut out	cut out	cut out	cut out	cut out	cut out	cut out	cut out
100	72	40	7	24	10	44	27	22	6
40	108	96	56	36	15	36	33	6	1
10	54	64	21	30	35	12	6	18	8
20	9	48	28	12	25	16	9	14	3
30	99	80	63	42	60	24	21	8	7
80	81	32	77	66	40	20	15	20	10
60	27	8	14	54	30	28	36	10	5
110	36	72	84	18	50	8	24	12	12
70	90	88	35	48	20	32	18	24	2
50	45	16	42	72	5	48	30	2	11
120	63	24	49	6	55	40	12	4	9
90	18	56	70	60	45	4	3	16	4

Write a number in the center of the Web, then multiply each of the numbers inside the web by that number. Write the answers around the outside.

time in sec. × # correct